P9-BHS-339

Klaus Dören/Werner Freitag/Dieter Stoye · Water-Borne Coatings

Klaus Dören / Werner Freitag /
Dieter Stoye

Water-Borne Coatings

The Environmentally-friendly Alternative

32 Tables / 21 Figures

Hanser Publishers, Munich Vienna New York

Hanser / Gardner Publications, Inc., Cincinnati

ɔy

44-4090, USA

Distributed in all other countries by
Carl Hanser Verlag
Postfach 860 420, 81631 München, Germany
Fax: +49 (89) 98 12 64

Library of Congress Cataloging-in-Publication Data

Dören, Klaus.
[Wasserlacke. English]
Water-borne coatings : the environmentally-friendly alternative / Klaus Dören. Werner Freitag, Dieter Stoye.
p. cm.
Includes bibliographical references and index.
ISBN 1-56990-139-2
1. Emulsion paint. I. Freitag. Werner. 1951- II. Stoye, Dieter. III. Title.
TP935.D6713 1994
667'.9--dc20 94-29398

Die Deutsche Bibliothek – CIP-Einheitsaufnahme

Dören, Klaus:
Water borne coatings : the environmentally friendly alternative ; 32 tables / Klaus Dören/Werner Freitag/Dieter Stoye. – Munich ; Vienna ; New York : Hanser ; Cincinnati : Hanser/Gardner, 1994
ISBN 3-446-17632-2 (Hanser) Pp.
ISBN 1-56990-139-2 (Hanser/Gardner) Pp.
NE: Freitag, Werner:; Stoye, Dieter:

Preface

Water-borne coatings are gaining steadily in significance as environmentally protective alternatives to solvent-containing coatings systems. The reason could be their liquid consistency; they are more comparable in their handling and processing with conventional, likewise liquid coatings systems. Their market acceptance is therefore much greater than that of other low-emission coatings.

We have therefore decided to make our contribution to this development by writing a book about the fundamentals, raw materials, developments and technologies of water-borne coatings. The basis for this is the lectures at the Technical Academy of Wuppertal, and in some cases, the seminars in Wuppertal, Nüremberg and Cottbus, Germany that we have presented. We were supported from many sides in this venture. On behalf of all, we wish particularly to thank Dr. Christina Machate, who wrote the contribution on quality assurance. Mr. Klaus Kock is also due to our thanks for his help. With his thorough knowledge of computing, he assisted significantly with organizing the graphics and lay-out. We thank the management of Hüls AG in Marl, Germany, for the technical support which was at our disposal and for permission to publish results that were developed in the laboratories and institutions of the company. We wish the reader of this book a deeper insight, through his studies, into the status of water-borne coatings as a basis for new perceptions and developments in this field.

Marl, Germany
Sommer 1994

Klaus Dören
Werner Freitag
Dieter Stoye

Table of Contents

1 Introduction

The progress of industrial development in most of the countries of the world has brought with it an increased pollution of the atmosphere with volatile organic substances. The causes are varied. Industrial production always results in concentrating and conglomerating traffic and industrial facilities as well as working and living people – tied to increased waste and effluent quantities, emission of dusts and organic products. There are two chief human sources of organic emissions:

- industrial production facilities, and
- residential and recreational activities.

The following industrial activities play a significant role in the escape of organic emissions:

- the coatings industry
- the chemical industry
- crude oil refining
- the printing and adhesives industries.

The emissions released annually by these areas in the European Community are shown in Table 1.

The emissions attributable to residential and recreational activities have their source primarily in

- heating, as well as
- individual and public transportation.

The main cause of all organic emissions from human sources is private vehicular traffic [1].

In this context however, it must not be forgotten that it is estimated that organic emissions from anthropogenic sources constitute approximately 15% of the global emission of organic substances; 85% are produced by nature through rotting and decay processes or in the framework of normal plant, animal and human metabolism.

With a contribution of approximately 20 – 25%, the coatings processing and manufacturing industry represents a quite significant source of organic emissions derived from human activities. It is in second place after vehicular traffic as a producer of atmospheric pollutants. Solvents from the coatings industry and other volatile products resulting from painting are considered to be organic emissions.

Because of the high contribution of solvents, and in the context of a growing concern for the environment, it is the prerogative of the coatings manufacturer, the coating producer and the coatings processing industry to reduce drastically the solvent content of coatings systems by any means possible [2]. This resulted in the fact that the coatings industry since then has concerned itself almost exclusively with the development and research of low-solvent and solvent-free binders and the coatings producers are mainly pursuing new developments based on this new product group. In addition, major coatings users such as the automotive and appliance industries are making corresponding demands on their paint suppliers to make low-solvent paints available. Furthermore, there is the increasing difficulty and expense of disposing of solvent-containing wastes such as coatings residues and sludges, which are considered hazardous waste. Today, new investments in paint facilities are always made against a background of low-solvent, and therefore more environment-friendly coatings, even when there may be no low-solvent alternatives available in the near term because of reasons of technical feasibility and high demands.

Table 1 Relative Emissions Sources for Organic Substances in the European Community

	Organic Emissions (%)
Traffic	53
Solvents	34
Others	13

Total Annual Quantity in Metric Tons: 10 Million

Before assuming the task of replacing the solvents present in conventional coatings, one should be clear about the mode of action of the solvents themselves. The solvents in coatings serve both as an aid in processing and to control and improve the quality of the coatings and their use. They assume the following missions:

- Dissolving the binder
- Homogenizing various binder components
- Improving the wetting of pigments and fillers, reducing pigment floating
- Increasing the coatings shelf life
- Standardizing the electrical conductivity of electrostatically processed coatings
- Regulating the processing viscosity
- Optimizing the processing of coatings in differing manufacturing configurations with minimal problems
- Favoring the wetting of substrate surfaces that may not always be grease- and dirt-free
- Directing the sag behavior of the coating on vertical surfaces
- Regulating the physical drying

- Improving the adhesion of the coating to the part being coated.
- Furthering desired surface effects and structures
- Regulating flow development and increasing gloss
- Reducing of cratering, blistering, runs and haze.

This is an abundance of assignments for solvents, which both describes the advantages of solvent-containing coatings and delineates the difficulty and effort of developing low-solvent and solvent-free coatings alternatives.

On the other hand, there are the well-known disadvantages associated with the use of solvents in coatings. Once the solvents have completed their "work" in coatings, they should escape from the coating as fast as possible; it is their goal to end up in the atmosphere. We know well that most organic solvents are rapidly degraded biologically. In heavily industrialized densely-settled countries and in industrial conglomerations however, the localized quantity of solvents that ends up in the atmosphere is sufficiently large that damaging effects of an ecological nature cannot be excluded. Finally, there is also a question of economics and raw materials sourcing in whether it makes sense to release large quantities of solvents from coatings and printing inks to the atmosphere unused, without first evaluating technically sensible alternatives.

2 Problems with Solvents in Coatings – Occupational Hygiene and Ecology

All organic solvents are toxicants for organisms in the concentrations in which they are employed in coatings. On human skin they act to remove fat, they remove the protective layer and thus indirectly further the formation of skin redness, rashes and even inflammation by enabling the external influence of atmospheric agents, chemical toxicants, bacteria or fungi. Inhaled solvent vapors affect in differing ways the circulatory system, the nervous system, the lungs and the liver depending on the type of solvent. Humans and the various animal species can thus react and be injured differently. Extrapolating the effect of toxicants from animal studies to their effect on humans is thus not always justified.

Never-the-less, it is certainly beneficial from an occupational hygiene perspective to reduce the quantities of solvents used in coatings and to replace them wherever possible. Ecologically, solvent vapors are implicated in the formation of smog and the die-off of forests. Discussions of whether this is substantiated have not yet come to a conclusion. To never-the-less limit the amount of organic solvents in the atmosphere, they have been defined as toxicants in different countries by, for example, the Clean Air Act in the United States and the regulation "Technical Guidelines for Preserving the Purity of the Air 1986 [TA Luft)" in Germany [3].

Upon entering the soil, most solvents are degraded by the microbacteria and fungi that are present, so that pollution of the ground water usually need not be feared. In the case of transportation accidents and in the case of solvents that degrade with difficulty, a detrimental effect on the ground water is however, a definite possibility.

The coatings wastes and residues formed by the manufacture and processing of coatings are hazardous waste and can only be stored in puncture-proof form in special repositories. Preferably however, they are incinerated in strict compliance with official and legal guidelines. The premise here is that water dilutable coatings residues are also to be handled as hazardous waste. The waste problem can in any case be partially solved by changing the coatings system, or preferably by developing new recycling processes (see Chapter 17.4).

For these reasons, it is in the interest of us all to concern ourselves with the development of low-solvent and solvent-free alternatives, even if there is as yet no consensus in the factual – and sometimes not factual – discussions about the extent of damage to our environment by solvents.

3 Alternatives to Solvent-Containing Coatings Systems

The desire to replace solvents in coatings is certainly not new. For many years already, water-reducible dispersion brushing paints have been used which have replaced solvent-containing products at a time when one did not yet speak of environmental protection. Also water-reducible industrial coatings. e.g. electrodeposition coatings in the automotive industry, were developed in the sixties, when the word ecology was still foreign. These developments resulted mostly for technical reasons, less from a consideration of occupational hygiene or environmental protection.

In past years, numerous alternatives have been developed as low-solvent and solvent-free coatings systems:

- water-soluble coatings
- aqueous dispersions
- low-solvent high-solids coatings
- powder coatings
- radiation-curing coatings.

Table 2 gives a statistical overview of the quantities and types of industrial coatings produced in 1990 (see also chapter 18, table 32)[4]. An annual growth rate of 6–8% is estimated for water-borne coatings for industrial use. Their share of the total coatings production is just 5% today (including electrodeposition coatings), while the share of emulsion paints including resin-bound plasters is 40%. Especially high is the growth rate of the industrial dispersion coatings with a still small share of the total coatings market (2–3%).

Table 2 Types of Industrial Paint Systems Worldwide in % (1990)

Coating Systems	%
Water-Borne Systems	35
High-Solids	18
2-Pack Systems	16
Powder Coatings	11
Radiation-Curable Coatings	5
Solvent-Containing Coating Systems	35

The water-soluble coatings [5] (see chapter 6) contain binders that, either because of their strongly polar structure are intrinsically soluble – or at least swell – in water, or which after chemical reaction, usually with the formation of binder salts, pass into the water-soluble state. The binders that are water soluble due to their structure remain in principle water soluble even after drying – a disadvantage, which must be compensated by combining with other binders or resins, through chemical reaction with hardeners or by supplemental protection of the coating.

Water-soluble binder salts can result from the reaction of binders having acid groups with amines, or the reaction of binders containing basic groups with acids. Volatile bases or acids should always be used as the neutralizing components. These are capable of escaping during the drying of the coating, with the reconstruction of the water-insoluble acidic or basic binder. Such binders can also be cross-linked at their reactive groups at room temperature or at elevated temperature, which further improves their water resistance properties. Stability to dampness and water is however, an important condition for corrosion protection coatings.

A prerequisite for a sufficient water solubility of both water-soluble binders and binder salts is a relatively low molecular weight of ca. 1000 to 6000, in exceptional cases up to 20,000. If the molecular weight of the binder is too high, the products are no longer water soluble but can only be dispersed. Between the water-soluble binders and the binder dispersions there are transitions, characterized by clear solutions with a specific optical refraction, the so-called Tyndall Effect.

Emulsion paints and coatings derived from water-soluble binders have in common the property of water reducibility. In contrast with water-soluble systems, emulsions consist of finely divided water-insoluble polymer particles with a molecular weight of approximately 30,000. With film building, the particles flow together creating an integral film. The prerequisite for an unobjectionable film building is a film-building temperature formulated for the ambient

Table 3: Differences in the Dispersity of Binders for Water-Borne Coatings

	Solution	Colloidal Solution	Emulsion
Appearance	Clear	Opalescent	Opaque
Tyndall Effect	None	Slight	Present
Particle Size	0.01 µm	< 0.1 µm	0.1 – 0.5 µm
Molecular Weight	< 20,000	< 200,000	> 200,000
Viscosity	High, Molecular Weight Dependent	Moderate, Molecular Weight Dependant	Low, Molecular Weight Independent
Solids Content	Low	Medium	High
Amine Content	Medium	Medium	Slight

drying temperature, which can be controlled in conjunction with film-building agents (see chapter 8.3). Never-the-less, it is the rule that coatings from emulsions do not exhibit the film integrity, at equal thicknesses, that can be achieved with dissolved binders. Of particular importance is also the anomalous difference between both water-borne systems, which occurs with pigmentation: While the binders dissolved in water wet the pigment on its surface upon pigmentation and coloration, thereby determining the brilliance, color depth and gloss; with emulsion paints, an inner mixing of the pigment granules and binder is precluded – pigment and binder lie next to each other in the dry film, resulting in a reduced brilliance and color depth, as well as a reduced surface gloss.

Water-soluble coatings and emulsion paints are comparable with conventional solvent-containing coatings in their viscosity and important processing properties. Often the same applications techniques can be used to effect a coatings conversion. Particular attention must be paid however, to coatings feed lines due to higher corrosion susceptibility, the viscosity stability in relation to the shear forces with pumping the coatings systems in circulation and the viscosity constancy with processing aqueous coatings with standardized thixotropy, as for example, structure coatings.

Low-solvent coatings, so-called medium-solids or high-solids coatings can be manufactured in two ways:

- Basing the selection of the solvent to be used not upon price, but upon the ability to dissolve the binder with low viscosity and high solids content
- Development of low molecular weight binders with better solubility in solvents.

Solvent selection can be empirical or based on calculations using the solubility parameters of the solvents and binders. The principle of solubility parameters will not be discussed further here [7].

Fundamentally, the solubility parameter represents a thermodynamic value which gives a quantitative prediction of the type and strength of molecular interactions between mixing partners – in this case binders and solvents. The molecular interactions influence the viscosity so that a prediction of the solution viscosity based on the solubility parameters is possible [8].

The development of low molecular weight binders can result from physically- and chemically-drying binders. In both cases, there are natural limits to the venture. The molecular weight of the physically-drying binder can not reach values that are too low, or the coatings may become susceptible to dirt or even tacky, as well as lose their mechanical properties of hardness and elasticity. In the case of chemically-drying binders, the functionality increases rapidly with decreasing molecular weight; the binders become very polar and moisture sensitive. In addition the use of large quantities of hardeners are required which often results in a strong cross-linking, associated with increased brittleness of the coating. In both cases, the search for an optimum compromise between technical properties on the one hand and low solvent content on the other is mandated. Powder coatings are the coatings systems with the lowest volatile substances content. The procedure consists of electrostatically spraying a finely-divided, usually pigmented powder on the grounded object to be coated. The following factors represent drawbacks:

- Danger of dust explosions
- High initial investment
- Difficult color change
- Only relatively thick films are possible
- Orange peel.

Today's powder spray facilities have reached high safety standards, which in conjunction with clean working conditions and corresponding purification expenditures assure in all respects good security against dust explosions. Flexible spray facilities with mobile booths for the individual colors permit color change in a relatively short time. Film thicknesses of 60 μm provide good protection; certain structural effects, for example, for household appliances are desired for optical reasons.

The technical properties of powder coatings, such as abrasion resistance, water resistance, chemical stability, and very high corrosion protection are quite excellent. These technical properties in combination with the observed environmental protection are the reasons for the above average worldwide growth in the use of powder coatings. There are two processes for powder coatings systems;

- Fluidized bed
- electrostatic powder coatings.

In the fluidized bed process, the articles to be coated are heated to about 360°C or higher and are submerged in an air-fluidized powder bed. The powder granules sinter on the hot surface of the article, where depending on the heat capacity of the material, they melt to form a coating, sometimes with auxiliary heating. This process is suited, depending on bed capacity, for small to medium sized articles. Dishwasher baskets, printing rolls, armatures, derricks, office scissors, pens, hooks and eyes and other objects with high corrosion protection, water resistance, chemical stability and abrasion resistance requirements, are coated by this process. Thermoplastic polymers , such as polyamide -11 and -12, have proven to be especially useful.

Among electrostatic powder coatings, the following have achieved preeminence:

- the hybrid system of epoxy resin and carboxylated polyester powder,
- the TGIC-system (carboxylated polyester with triglycidylisocyanurate as cross-linker)
- the IPDI-system (hydroxylated polyester with blocked isophorone diisocyanate as cross-linker).

There are two groups of radiation-curing coatings systems suited for use in the "niche realm":

- Coatings systems of a reactive polymer, dissolved in a reactive diluent. Such polymers can be unsaturated polyesters or polyacrylates with unsaturated groups. Acrylic esters of low volatility are suited as reactive diluents.
- Coatings systems of a radiation-reactive polymer dispersed in water.

Table 4 Comparison of Coatings Systems

	Water-Borne Coating	Powder Coating	High-Solids Coating	Radiation Curing Coating	Solvent Containing Coating
Environment Protection	+	++	0/+	++	−−
Pigmentation	0	−	+	−	++
Special Coatings Effects	0	−	+	−	++
Application	+	0	+	0	++
Reactivity	0/−	−	−	+	++
Energy Balance	0	+	0	+	++
Raw Materials Requirement	+	++	0	0	−
Investment Level	0	−	+	−−	+

Weights: ++ very good, −− unsatisfactory

UV and electron beam are suitable radiation sources. The primary application areas are the printing ink industry, the paper and cardboard printing industry and the wood industry.

A comparison of the low-solvent systems that have been described, with conventional solvent-containing systems in terms of some environmental, raw material, any technically relevant criteria is shown in table 4.

As can be seen in table 4, the conventional solvent-containing coatings are naturally the most developed in terms of technical properties and processing. With powder coatings and radiation-curing systems, the positives and negatives are especially extreme: a typical evaluation for techniques that cannot be used universally, but in many areas have extraordinary advantages over conventional standards. Water-borne coatings and high-solids coatings, as liquid systems, are the closest to solvent-containing systems, with advantages in environmental protection, raw materials and energy requirements, but with weaknesses in the usual coatings demands.

It is the purpose of the present publication about water-borne coatings to, if not eliminate, at least reduce the weaknesses of this product group.

4 Historical Foundations of Water-Borne Coatings

For millennia, man has been concerned with the beautification of his environment by paintings, by dyeing his clothing, by coloring his skin and face, or by painting and inscribing walls for aesthetic reasons or for the information of his fellow human beings. The first painters used mineral colors, which they suspended in water or set with water and albumin. Fats, talc and animal and plant oils were also used as binders. Water-reducible and low-solvent coatings existed at the beginning of the development of coatings. The use of solvents became known only about 300 years ago. They were used to dissolve solid natural resins that gave fast-drying paints.

The real development of coatings solvents did not begin until our century. It is tied to the assembly line for rapid production of mass-produced products. Assembly line production required the use of quick-drying paints, which became available in solvents at the beginning of this century from solutions of nitrocellulose – a byproduct of the production of gun cotton during World War I. The most important solvent patents stem from the twenties of this century – the patents of the large chemical companies, which are as important now as before. It was a revolutionary step to use solvents, without which industrial development and our present standard of living would be inconceivable.

Today we again stand at the threshold of new developments that have the goal of preserving our standard of living and at the same time of protecting us and our environment from the damages of industrialization which have nicked at our well being. This can only imply that a reversal is neither desirable nor sensible. Only the further development of technology, which in our case encompasses a further development of coatings techniques, can offer a solution [10].

5 Water – A Fundamental Constituent of Water-Borne Coatings

In water-borne coatings, water serves as solvent and dispersion medium. Compared with the usual solvents, water has decidedly different properties (see table 5):

- Water freezes at 0°C. As a rule, water-borne coatings should therefore be stored above the freezing point. It must be checked in every case, whether the coatings technical properties (stability, processing, surface properties) are changed by the freezing process.

Table 5 Comparison of the Properties of Water and Solvents

	Water	Organic Solvent (Xylene)
Boiling Point °C	100.0	144.0
Freezing Point °C	0.0	–25.0
Solubility Parameter $(J/cm^3)^{1/2}$ acc. to Hansen δ_D δ_P δ_H global	 12.6 32.1 35.1 49.3	 17.8 1.0 3.1 18.0
Hydrogen Bonding Index	39.0	4.5
Dipole Moment Db	1.8	0.4
Surface Tension σ mN/m	73.0	30.0
Viscosity mPa s	1.0	0.8
Relative Volatility (Diethyl Ether = 1)	80.0	14.0
Vapor Pressure 25°C, hPa	23.8	7
Specific Heat J/g deg	4.2	1.7
Heat of Vaporization J/g	2,300	390.0
Dielectric Constant	78.0	2.4
Thermal Conductivity x 10^3 W/m^2 deg	5.8	1.6
Relative Density d_4^{20}	1.0	0.9
Index of Refraction n_D^{20}	1.3	1.5
Flash Point °C	–	23
Lower Explosive Limit, Vol %	–	1.1

- Water boils at 100°C and vaporizes as a homogeneous substance with a relatively low relative volatility compared with solvents. In solvent-containing coatings, on the other hand, solvent mixtures are present that provide for a time-wise extended, uniform vaporization and therefore the development of a smooth coatings surface. In the case of water-borne coatings, difficulties with the optical surface quality of the coating must be anticipated for this reason. To counter these, auxiliary solvents and film-building agents are used in water-borne solvents (see chapters 8 and 10). At the same time, the use of additives and the correct selection of binder must contribute to mastering surface problems.

- Water has a decidedly higher surface tension than organic solvents. This leads to a poorer wetting of the substrate to be coated. In working with water-borne coatings, it is necessary to provide for a good cleaning of the substrate. The binders developed for water-borne coatings can contribute much to overcoming the physical problems associated with the "product water". Only seldom however, does the binder bring all the advantageous properties that the organic solvent provided "for free" before. Additions of auxiliary solvents serve to reduce the surface tension of the water (see chapters 8 and 10).

- Water has a higher enthalpy of vaporization compared with solvents. The drying of water-born coatings therefore requires the addition of greater amounts of energy and, as a rule, also requires more time (see chapters 14 and 15). This plays no decisive role when coating absorbent substrates, since the water is drawn into the substrate. In other cases, the higher energy inputs required for drying are compensated for by the fact that ventilation can be reduced in processing water-borne coatings compared with solvent-containing coatings. With solvent-containing coatings, ventilation must always be kept high for safety reasons, because of the danger of explosion, which is to be feared when the explosive limits of the solvent-air mixture are exceeded. With water-borne coatings, on the other hand, it is sufficient to limit ventilation to the amount necessary to remove, in addition to the non-flammable water vapor, the auxiliary solvents and condensation products, if any. A reduction in ventilation is coupled with savings in energy. A further energy saving is achieved by eliminating or reducing the need for a thermal solvent after-burner, whose operation is energy intensive. Fundamentally, one can assume that water-borne coatings can be applied for about the same energy costs as solvent-containing coatings (see chapter 15). Under certain circumstances, as much as 30% of energy costs can be saved, as shown by a comparison of water-reducible baking coatings with solvent-containing coatings [11].

- Water is not flammable, an advantage that can affect the cost of insurance premiums for fire insurance, and is also of significance in terms of handling, storage and shipping.

- Water has totally different solubility parameters than organic solvents. It is definitely polar and forms significantly stronger hydrogen bonds, than those known for organic solvents. As a result, the interactions between binder and water molecules are also of a different quality and quantity than in the case of solvent-containing coatings. This impacts the selection of binders for water-borne coatings, which must approach water itself in their polarity properties and from which one expects strong hydrogen bonding in the system as a whole. This behavior affects the viscosity of water-borne coatings, although the viscosity of water is completely comparable with that of organic solvents. The polarity and hydrogen bonding tendency of water also influence the dilution behavior of the coatings and the wetting of other less polar partners, such as pigments, fillers or also substrates such as plastics, paper or metal surfaces. Further, the ability to absorb dirt and grease, which constitutes the broad application tolerance of solvent-containing coatings systems, is greatly reduced by the characteristics of water that have been described.

- The differing values of water for the dipole moment and dielectric constant, compared with those for organic solvents, lead one to expect the properties for water-borne coatings that were already mentioned for the solubility parameter differences.

- The electrical and thermal conductivity of water are of a different magnitude than those of organic solvents. In contrast with solvents, water used in coatings is not an insulator. This is the reason for the problems to be anticipated in conceptualizing electrostatically sprayed water-borne coatings (see chapter 13.2).

6 Types of Aqueous Coatings Systems

6.1 General Considerations

The aqueous coatings systems can be divided according to criteria such as:

- Type of coatings binder
- Means of drying or
- Applications area.

The most sensible approach is to consider the different binders according to the nature of their stabilization in the aqueous phase. One differentiates among "true solutions", colloidal solutions, dispersions [12] and emulsions (see also table 3).

Table 6 Typical Product Properties of Aqueous Coating Systems

Property	Solution	Colloidal Solution	Primary Dispersion	Secondary Emulsion
Appearance	clear	Bluish Translucent	Opaque	Milky Opaque
Molecular Weight	< 20,000	< 100,000	> 100,000	< 20,000
Particle Size	< 0.01 µm	< 0.1 µm	> 0.1 µm	> 0.1 µm
Viscosity	High	Medium	Low	Low
Solids Content	Relatively Low	Medium	High	High
Solvent Content	to 25%	to 10%	< 10%	< 10%
Pigment Wetting	Good	Good	Poor	Poor
Flow	Good	Medium	Poor	Poor
Applications Latitude	Great	Limited	Limited	Small

Table 6 gives an overview of typical product properties. All systems have advantages and also disadvantages. In the English and American literature one frequently finds the classification

Solution	0.001 µm	(water soluble)
Colloid	0.001 – 0.1 µm	(colloidal dispersion)
Dispersion	> 0.1 µm	(aqueous emulsion),

which partially undercuts the concepts.

The organization used in table 6 presents more possibilities for arriving at a conclusion about differing application possibilities based upon particle size and, as the case may be, the production process of the aqueous binder.

Combinations of binders are often utilized to unite the advantageous properties of different types of aqueous binders, e.g. solutions and dispersions. These coatings are designated as hybrid systems.

As a clarification, the dependance of particle size on the production process is shown graphically in figure 1 [14].

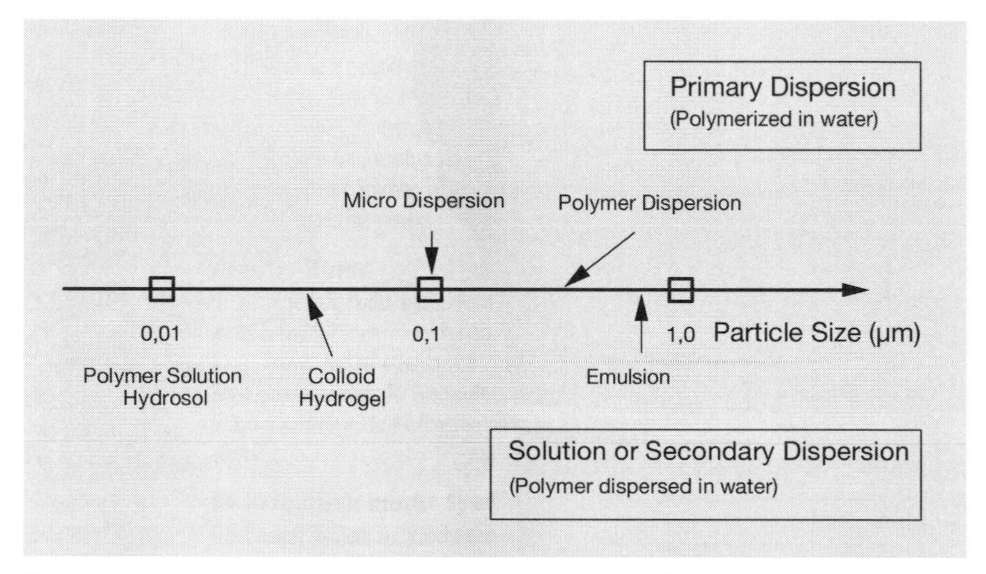

Figure 1 Particle Size Dependance on Manufacturing Process

In ideal cases, the binder solutions or water-soluble binders can exist as separate molecules dissolved in water (homogeneous mixture). In actual practice, complicated structures are usually encountered which are strongly dependant on the degree of dilution and further factors, such as temperature, added solvent, etc.

Water-soluble binders are divided into non-ionic, anionic and cationic products:

A non-ionic water-soluble binders

- Collagen
- Cellulose esters, -ethers
- Polyglycolethers
- Amino resins
- Polyvinyl alcohols
- Polyvinylmethyl ethers
- Styrene-allyl alcohol-copolymers

B anionic water-soluble binders

- Water glass
- Alkyd resins
- Saturated Polyesters
- Phenolic resins
- Acidic Polyacrylates
- Maleic acid-copolymers
- Polybutadiene-maleic anhydride-adducts

C cationic water-soluble binders

- Amine-modified epoxy resins
- Amine-modified polybutadienes

The binders of group A are soluble in water solely due to their chemical make-up, i.e. because of the abundance of polar structures in the macromolecule. In general, alcohol, ether, and amino groups are decisive. These resins are considered in chapter 7 (Binders). Groups B and C become water soluble by salt formation across functional groups. These are discussed in greater detail in chapter 6.2. (see also Lit.[15]).

Dispersions, on the other hand, are two-phase systems that are not water-soluble, but only water-reducible. They are produced by polymerizing emulsified monomers in water. The following monomers are frequently used:

- Acrylates
- Methacrylates
- Styrene
- Vinyl acetate
- Ethylene
- Butadiene
- Vinyl chloride
- Other vinyl esters (see chapter 7.12.)

In addition to homopolymerizates, copolymers are above all of importance, since many properties can be targeted by selection of raw materials [17]. Micro dispersions are considered to be especially finely divided polymer dispersions that can be obtained by special techniques and by using special emulsifying agents. They are in the medium particle size range of about 0.1 μm. Even finer namely in the range of 0.01 to 0.1 μm, are the colloids and hydrogels. Like the water-soluble resins, they are usually first bulk polymerized or polycondensed, and subsequently dissolved in water over salt-building groups and incorporated ether groups. They are not especially differentiated from the resin emulsions, which include all resins emulsified after the polymerization process (secondary dispersions), independent of whether additional, not chemically bound emulsifying agents are employed.

Commercially important resin emulsions are:

- Alkyd resin emulsions
- Epoxy resin emulsions
- Silicon resin emulsions
- Bitumen emulsions.

Of these, the modified alkyd resins and epoxy resins are especially significant with the hydrogels [18]. The various forms of dispersions and emulsions possess the common feature of two-phase construction. They are characterized by a dispersed and a continuous phase (solid-liquid or liquid-liquid). Stability exists only when the thermal energy of the particles is less than the sum of the energy of electrostatic repulsion and van der Waals attraction. To avoid coagulation of the binder and aqueous coatings system, this delicate equilibrium must not be disturbed. The following factors that can effect coagulation must be considered:

- Acids
- Salts
- Water-soluble substances, e.g. solvents
- Colloids with opposite charge
- Heat
- Freezing
- High shear forces or pressures
- Evaporation of water
- Electrical current.

This multitude of factors should not give the impression that modern water borne coatings binders are extremely sensitive. Never-the-less, these influences are to be considered above all in coatings development and also in changing processing conditions. Above all, the stability during the desired processing time period is important.

To obtain good coatings stability properties, the binders must pass to a waterproof state after coatings application. In general, the same principles are used as with solvent-based coatings, as e.g.:

- physical drying:
  evaporation of water, amines (anionic binders) or acids (cationic binders); in some cases, auxiliary solvents

- oxidative drying:
  additional cross-linking by oxygen

- thermal cross-linking (baking enamels):
  thermally induced cross-linking through self condensation across suitable functional groups or condensation with cross-linking resins.

Coatings materials exist as

- one-component coatings, i.e. one or more binder components in the coatings formulation with sufficient stability under normal storage conditions or as
- two- or more component coatings, i.e. two or more binder components are mixed immediately before processing in the form of the finished formulation, due to the limited storage time (pot life) of the total system.

6.2 Water-Soluble Systems

Water-soluble binders can exist in the aqueous phase in nonionic or polyionic form (see chapter 6.1.). While the ionic binders usually have resin-like characteristics with molecular weights below Mn = 20,000 g/mol, the nonionic binders are often high molecular. Examples are polyvinyl alcohols, ethylene-maleic anhydride-copolymers and polyacrylamides, which are however seldom used as coatings binders. The polyethylene oxides (polyethylene glycols), with molecular weights from 200 to the high molecular weight range, are more significant as thickeners and for chemical modification. Of greater interest to the coatings practitioner are the phenolic and amino cross-linking resins (see chapter 7).

By far the largest portion of water-soluble binders for the coatings sector consists of ionic resins, i.e. water solubility is achieved by neutralization or salt building. The resin can be positively (cationic) or negatively (anionic) charged (formulas see below). Cationic resins, as a rule, are modified with amines and therefore can be protonated with acids. Of greatest significance are amine-modified epoxy resins and polybutadienes, which are used above all in electrodip coatings (see chapter 13.2.6. and 16.1.1.)

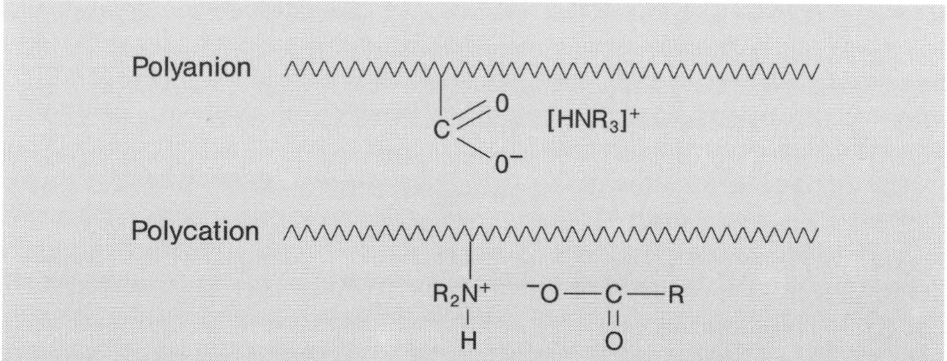

Polyanion: $-C(=O)O^-$ $[HNR_3]^+$

Polycation: R_2N^+H $^-O-C(=O)-R$

The anionic resins frequently possess organic acid groups which are neutralized with amines or ammonia (acid number of the resin > 40 mg KOH/g).

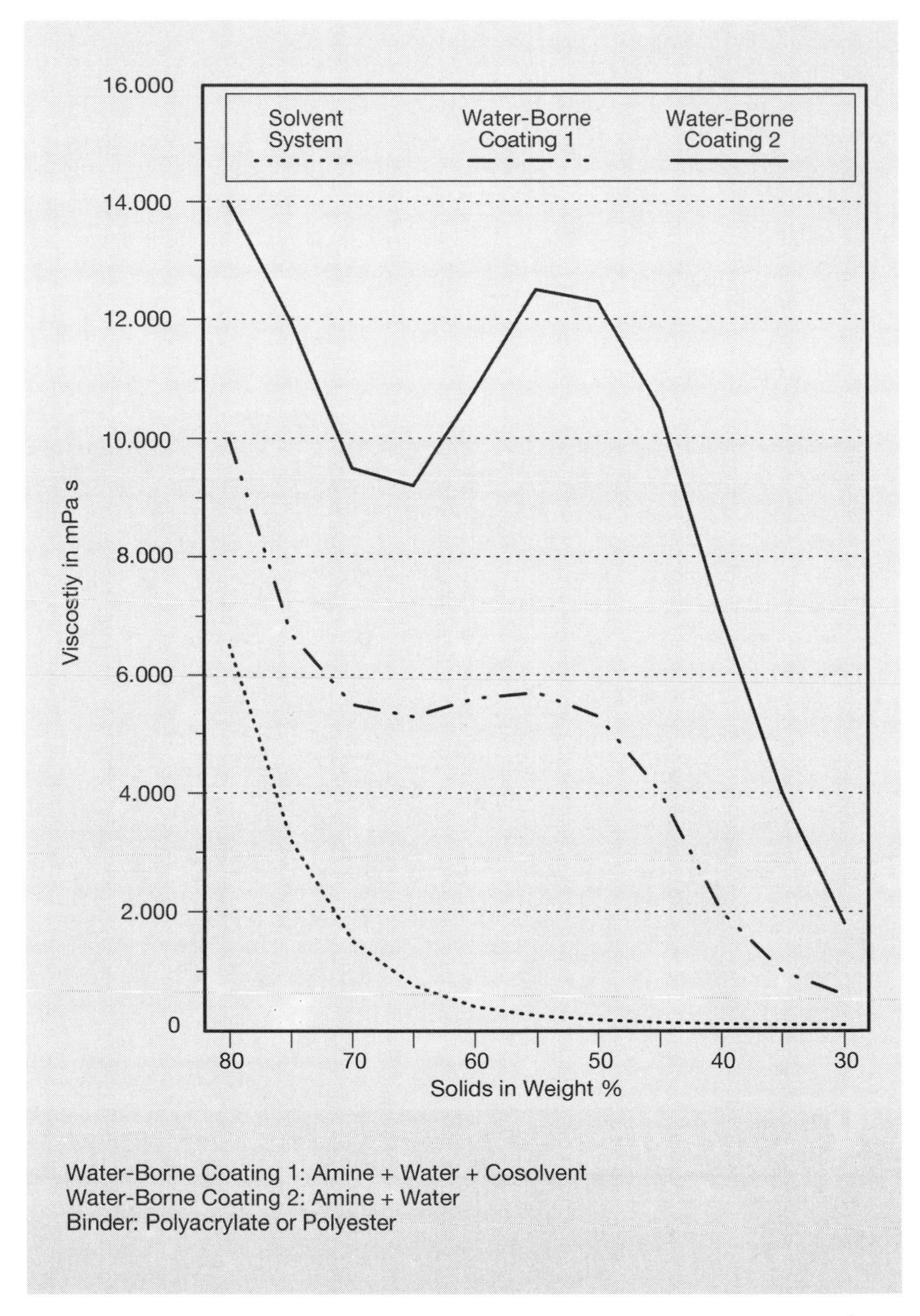

Figure 2 Viscosity of Water-Borne Coatings in Relation to Composition in Comparison with Solvent-Containing Coatings.

In practice, alkyd resins, saturated polyesters and epoxy resin esters are encountered most. These are the "classic" binders for water-borne resins whose properties have been significantly improved in recent years.

The drawbacks of the water-soluble systems are based on the fact that auxiliaries are required to obtain the aqueous binder solution. The previously mentioned neutralizing agents (acids or amines) must escape during the drying process. They are thus involved in the ecological balance. Since heat is released during the neutralization process (exotherm), and the binders are often extremely viscous in solvent-free form, certain quantities of solvents (auxiliary solvents or co-solvents) must be incorporated, which represent volatile components at film building. (see chapter 8 and 10). The water-soluble binder system thus consists of the components

- acidic or basic polymer
- auxiliary solvent
- neutralizing agent
- water.

The components are usually mixed in this order. The quantities and chemical structure of the auxiliary solvent and neutralizing agent must be closely matched with the binder under consideration, since definite negative effects can result, especially when adjusting the viscosity with water (dilution behavior). The frequently observed interim increase in viscosity upon dilution with water is designated as a "water peak" or viscosity anomaly (see chapter 10).

Modern binder developments are distinguished by their small auxiliary solvent requirement (< 10%) and favorable dilution behavior [19].

The disadvantages of water-soluble systems are mostly confronted with their broad applicability under differing drying conditions and the multifaceted applications areas. To be mentioned are also the possibilities for producing high gloss coatings and coatings systems with high corrosion resistance. In summary, the properties of the painting are the most comparable with those of the solvent coatings.

6.3 Colloidal Solutions

The colloidal systems and also hydrogels have a series of things in common with water-soluble systems. Thus the binders, as a rule, are also bulk polymerized (usually by polycondensation). They also have neutralizable groups, which are converted by salt building with the production of the water-soluble system. To be sure, the number of these mostly acidic functional groups is not large enough to by themselves impart water-solubility (acid number of the solid resin under 35 mg KOH/g). Improved water solubility is achieved by the additional incorporation of hydrophilic groups in the polymer molecule. Polyglycols are the most frequently polymerized in. The resin thus receives self-emulsifying properties. The characteristics shown in table 6 result. The lightly milky appearing colloids consequently fall between the "true" solutions and polymer dispersions. Today alkyds and epoxy esters are the resins principally encountered in this form. The quantities of auxiliary solvents and amines (preferably ammonia) lie under 5 weight percent. A viscosity anomaly (water peak) on dilution with water is no longer encountered with modern products. These both air- and oven-drying systems are finding increasing distribution especially in industrial painting. Excellent corrosion protection as well as good drying and high gloss are attainable.

6.4 Dispersions

The characteristic criterion for polymer dispersions is their production in a polymerization process (usually by radical polymerization) in the presence of water. The formation of the polymer proceeds in emulsifier micella, with monomer droplets representing the raw materials source. Stabilization can result not only through emulsifying agents, but also through protective colloids. Dispersions of high polymers result, whose molecular weight no longer has an influence on the binder in its dispersed form, e.g. on the viscosity. Properties can be controlled by co-polymerizing varying monomers; functional groups can be incorporated. Modern preparations techniques such as core-shell polymerization are used to obtain optimal property combinations, for example of processing qualities and weathering stability (aqueous microgels [20]).

The classification of dispersions can follow multifaceted criteria. From a chemical synthesis perspective, vinyl acetate copolymerizates, styrene-acrylate-, pure acrylate- and styrene-butadiene-dispersions play the largest role for the coatings and paint sector. The wide offering of acrylate- and methacrylate monomers makes the range of copolymers barely comprehensible. A further classification can be based on the type and the electrical charge of the dispersing agent used (emulsifying agent, protective colloid) [16]. The quantity of dispersing agent affects the water stability and corrosion resistance of the paint films, since they remain in the film for an extended period. Also the size and homogeneity (monodispersity or polydispersity) of the polymer particles in the dispersion is important. Particles as uniform as possible of small size (microdispersion) or precisely formulated polydispersed systems, enable an undisturbed flow (coalescence) of the polymer on film building. An important variable of the dispersion is the so-called minimum film-forming temperature (MFFT). It predicts at what temperature an unobjectionable film building may still be anticipated. The MFFT is closely bound to the polymer synthesis and glass transition temperature. To lower the MFFT, coalescing agents and film-building additives are used (see chapter 8.3.). These mask suitable hard to volatilize solvents whose quantity in the paint usually lies below 5 weight percent.

Next to the advantages of dispersions, such as high solids content at the processing viscosity, small quantity of solvent and quick physical drying (water release), there exist a few application limitations. Thus it is difficult to achieve high gloss coatings or high corrosion resistance. Also the quick drying is disturbing, for example, with many spray and dip coatings. Modern developments are increasingly overcoming these problems. Besides, practically all binder classes in use can be combined with each other, so that specific properties can be targeted for optimization.

6.5 Emulsions

Designated as emulsions are those resins that are secondarily transferred to the aqueous phase, whose molecular weight still lies in the oligomer range and which display relatively large particle size. Most emulsions in the resin area are stabilized by chemically unbound emulsifying agents (mostly anionic or nonionic).

Typical resin emulsion [21]:	Parts
Resin	60
Emulsifying Agent	6
Water (Fungicide, defoamer)	34

The so-called HLB value (Hydrophilic-lipophilic-balance) is important for selecting the emulsifying agent [21]. It must be adjusted to the resin.

The resin emulsifiers can be produced quite cost effectively. Their particle size and the high emulsifier content (ca. 10%, based on the resin) often significantly reduce their applicability. Thus, gloss development and flow of the corresponding paints is often inadequate. Combinations with acrylate dispersions are therefore favored. The solvent content of emulsion-based paints can approach zero. Therefore there are intensive efforts to improve the above mentioned properties. Thus, incorporation of the emulsifiers with resin synthesis or at drying is being attempted. Also there are already examples of emulsifier-free dispersions. Here the boundary between emulsions and colloids is fluid. In addition to the resin emulsions already enumerated, there already exist promising prototype acrylate-secondary dispersions in the market. Thereby it has become possible to produce water-reducible industrial coatings systems from corrosion-protecting primers to top coats based on acrylates.

7 Binders in Water-Borne Coatings

In the preceding chapter, aqueous coatings systems were characterized above all according to the nature and type of their stabilization in the aqueous phase. While some monomer and binder classes have already been mentioned, we will now treat the individual coatings resins in greater depth [22]. After a short description of the chemical components of the binders and their drying-hardening mechanism, the form in which they are currently available in the market and the significance assigned to them will be described (see also the original literature in chapter 21).

7.1 Alkyd Resins

Alkyd resins are relatively low molecular weight polycondensates of multivalent alcohols and organic acids. In contrast to the (oil-free) polyesters, they always contain vegetable or synthetic, usually unsaturated, fatty acids (long chain monocarboxylic acids). According to the oil or fatty acid content, one differentiates

- short oil (to 40%)
- medium oil (40 – 60%) and
- long oil (more than 60%) alkyd resins.

They are further divided into air and oven drying types. Air drying proceeds, as a rule, by oxidation of unsaturated fatty acids and consequent cross-linking. There exists a broad range of possibilities for further chemical modification. Since in principle all polyesters exhibit the danger of saponifying their ester linkages in aqueous milieu, conversions that increase the saponification stability, such as e.g. acrylic and urethane modification, are especially carried out [23].

Since alkyd resins are always poly-condensed in bulk, the subsequent transfer into the aqueous form is mandated. All possibilities named in chapter 6 are available [18]:

a) Conventional alkyd resins can be brought into emulsion with the help of emulsifying agents. In so doing, the smallest possible particles and a narrow particle size distribution must be sought to attain satisfactory coatings technical properties. Even modern emulsions have problems with paint pigmentation, drying and gloss development, so that there is a range of limitations on the use of these products. Good results exist with wood impregnation and glazing, corrosion-protecting primers and nicotine resistant wall paints. Since relatively inexpensive resins without special modifications can be employed, further improvements, as e.g. the addition of cross-linkable emulsifiers are possible.

b) Typical "water-soluble" alkyd resins exhibit an acid number of over 40 mg KOH/g and are converted to the water-soluble salt form with volatile bases. Usually relatively large quantities of auxiliary solvents, as e.g. butylglycol, are needed to achieve good water-reducibility and storage stability [24]. The possible variations in selecting raw materials is very broad today, so that modern resins are saponification resistant and therefore are sufficiently storage stable for coatings formulation. Previously, the water-soluble alkyd resins were usually not pre-neutralized, but were supplied in water-reducible solvents (1st generation). In the paint factory, one was thus compelled to carry out the neutralization on one's own and to tolerate relatively high quantities of solvents in the paint [25, 26]. Today, many resins (2nd. generation) are already pre-neutralized and diluted with water. For the coatings producer, the problems encountered with dilution (water peak) are eliminated and also the total solvent content of the coatings material can be reduced by this measure. Often, to be sure, an amine addition is necessary, which under certain regulations is frequently assigned to a critical class. These resins are on the market for all known applications.

c) Designated as so-called 3rd. generation "water-soluble" alkyd resins are products that do not represent hydrosols, but hydrogels or colloids. These resins, which are also characterized as self-emulsifying, contain hydrophilic groups (e.g. polyglycols) and exhibit relatively low acid numbers (often under 20 mg KOH/g). Here too, there is no water peak upon dilution. Ammonia is usually used as the neutralizing agent, which under the different clean air acts is assigned to a less critical class.
Coatings materials of this basis, contain in addition to water, less than 5% volatile components (solvents and ammonia). Drawbacks are the often low solids content of the aqueous resin and their relatively high price, which results from the urethane and acrylic modification.

While the emulsions described under a) are still largely limited in their use, the resins of groups b) and c) are suitable for practically all applications areas known for conventional solvent coatings. Applications are known above all from the areas of industrial paints for machinery and vehicles, fillers, sealants and wood paints. A further increase in market share of these resins is anticipated.

7.2 Amino Resins

Amino resins are typical cross-linkers for all types of baking coatings. They are known for example, in the form of melamine, urea, or guanamine resins. Of these, the melamine-formaldehyde-condensates play the biggest role. Water-soluble types are obtained with a relatively low degree of polymerization and the use of short chain alcohols (methanol, ethanol) for forming ethers from the methylol groups. Incorporating carboxyl groups to obtain water solubility is also possible. A major role is played by hexamethoxymethylmelamine (HMMM), which is fully water soluble [27, 28, 29].

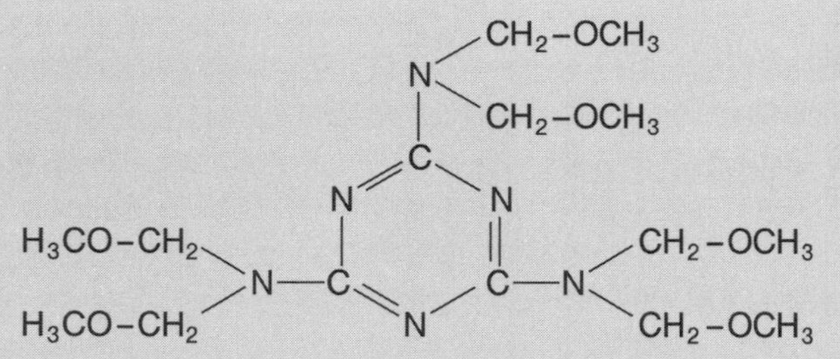

The amino resins can also be used with plasticizer binders (see also phenolic resins chapter 7.9) or brought in for modification with the synthesis of these binders. In both cases, heat cross-linking resins for baking coatings are obtained. The hardening proceeds by self condensation of the amino resin and cross-linking over hydroxyl or carboxyl groups of the combinations partner (see chapter 10). Typical baking systems are based on alkyd- or polyester-melamine resin- or acrylate-melamine resin-combinations. Good properties in regards to gloss retention, non-yellowing, and tint stability make these systems especially suited for decorative top coatings. Amino resin cross-linked baking systems play a large role in metal coating. The range of properties of the above named combinations based on water corresponds broadly with the level of conventional paints.

7.3 Cellulose Derivatives

Water-soluble cellulose derivatives are obtained primarily by esterification and etherization of cellulose. Familiar compounds are methyl-, carboxymethyl-, hydroxyethyl- and hydroxypropylcellulose. Their significance in the field of water-borne coatings lies above all in their role as protective colloids and thickeners. They are therefore treated with the additives [16] (see chapter 8).

In the area of solvent-containing coatings, especially wood coatings, nitrocellulose, now as before, is a key binder. In addition to the favorable price, a good property spectrum can be obtained through combination with other resins. In the United States especially, there are efforts to obtain aqueous systems [30].

Previous developments for water-borne coatings consist of nitrocellulose, alkyd resin, plasticizer and emulsifiers, for example. They are prepared in emulsion form and still contain 400 to 300 g/l volatile organic compounds (VOC). The nitrocellulose is utilized in water-wet form. The coatings properties obtained fulfill the special requirements of the wood industry (ability to wax, alcohol resistance, etc.) significantly better than other binders. The optical quality is on the same level as that of solvent-based nitrocellulose coatings. It remains to be seen how far the solvent content can be reduced.

The cellulose derivatives such as methylcellulose, hydroxyethylcellulose and sodium carboxymethylcellulose have a further true binder function in the so-called size colors. The cost-effective wall and trade paints have not lost their significance despite the great competition by products based on dispersions [16].

7.4 Epoxy Resins

Epoxy resins are obtained from epichlorohydrin and bisphenol A or similar monomers by condensation. Hydroxyl and terminal epoxy groups are present as reactive groups:

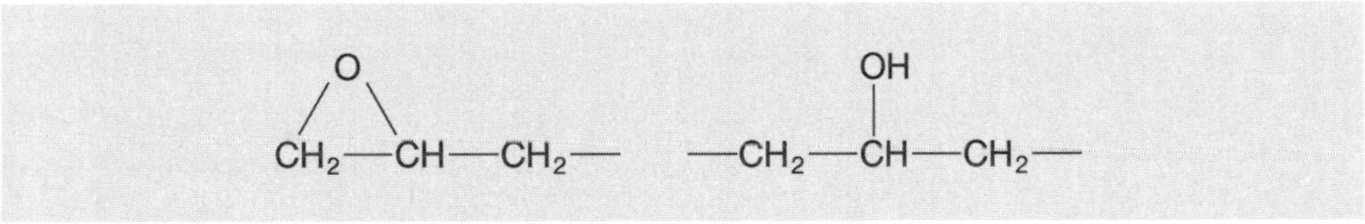

$$\underbrace{CH_2{-}CH}_{\text{O bridge}}{-}CH_2{-} \qquad {-}CH_2{-}\overset{OH}{\overset{|}{C}H}{-}CH_2{-}$$

Esterification with fatty acids leads to epoxy resin esters (see chapter 7.5). In addition, one can treat with amines or polyamides, which is used for both modifying and hardening [31].

Water-reducible epoxy resins are available in the form of

- liquid resins with emulsifier added
- liquid resins with chemically bound emulsifiers
- epoxy resins with reactive diluent
- emulsifier-containing dispersions, and as
- cationic resins for electrodip coating.

The first three mentioned are emulsified by the processor. Like the already aqueous dispersions delivered by the manufacturer, they must be reacted with suitable hardeners (mostly easily emulsifiable polyamines or polyamidoamines) in use [32]. These typical two-component systems are universally used where their good adhesion, toughness and chemical resistance are required. Thus high-value concrete sealants, floor paints and wall coatings are known.

Cataphoretically-precipitated electrodip paints were developed at the end of the seventies. Since then they have generally penetrated primer coats for automobile chassis [33). Most often used are the reaction products of epoxy resins and modified amines (for example various hydroxylamines). The amine adduct is protonated with acetic acid and is water-soluble in this salt form.

Cross-linking proceeds via. hydroxyl groups, that can originate from the side chains of the amine, by means of blocked isocyanates at baking temperatures of about 180°C. (see chapter 13 and 16).

$$-\underset{\underset{R_1\quad R_2}{N}}{\overset{|}{C}}-\overset{OH}{\underset{|}{C}}- \; + \; CH_3COOH \longrightarrow -\underset{\underset{R_1\quad R_2}{^{+}NH}}{\overset{|}{C}}-\overset{OH}{\underset{|}{C}}- \quad CH_3COO^-$$

7.5 Epoxyester Resins

Epoxyester resins are based in most cases on epichlorohydrin-bisphenol A resins that are esterified with fatty acids. The principles for obtaining water-soluble or water-reducible types are similar to those of the alkyd resins. Thus acidic components (e.g. maleic acid) can be reacted with hydroxyl groups, or one modifies the fatty acid entities, e.g. by copolymerization with acrylic monomers, including acrylic acid. Reactions with phenolic resins are also used.

$$\overset{O}{CH_2-CH}-CH_2- \; + \; R\text{-}COOH \longrightarrow \underset{\underset{\underset{R}{|}}{\underset{C=O}{|}}}{\underset{O}{|}}{\overset{OH}{CH_2}}-CH-CH_2-$$

The production of epoxyester resins is always carried out in a resin melt, so that subsequent transfer to water is needed. All of the methods listed in chapter 6 can be used. The historical development proceeded quite simultaneously with the alkyd resins, so that modern systems are already supplied pre-neutralized in water. Only small quantities of solvents are necessary [26].

Aqueous epoxyester resins are not suited for exterior trade paints with highly decorative requirements. Their forte is much more in both air and oven dried applications for corrosion protection and industrial priming. Fast-drying machine paints can also be formulated. Advantages are the good adhesion on nonferrous metals, the toughness and the good chemical resistance of the films.

7.6 Saturated Polyesters

Saturated polyesters consist of monomeric polycarboxylic acids and polyalcohols and are practically free of fatty acids. The molecular weight spectrum ranges from less then 1000 g/mol to over 20,000 g/mol, of which only low molecular weight types are used for water-borne coatings. For technical reasons, the resins are never used for physical drying. They are cross-linked externally, preferably with amino resins or polyisocyanates. For this, reactive groups such as carboxyl and hydroxyl groups are available [34, 35, 36].

Since the resins have relatively high melt viscosities and their structure allows only limited possibilities to attain water compatibility, exclusively acid polyester resins, which become water soluble through neutralization with bases and require certain portions of auxiliary solvents, are presently available in the market (see also chapter 7.1 and 10).

Water-soluble saturated polyesters are used in industrial baking paints, e.g. in combination with melamine resins. They display good properties relative to elasticity, surface hardness, gloss and weathering resistance and often parallel corresponding short-oil alkyd resins. Applications, e.g. in coil coating and office equipment, as well as automotive fillers and in the automobile supply industry are known. The paints are easy to process and lie below 10% in their amine and solvent content. The attainable property range of the water-borne coatings corresponds broadly with that of conventional coatings [37, 38, 39, 40, 41, 42].

7.7 Modified Oils

Vegetable and animal oils represent classical coatings binders. Triglycerides of unsaturated C-18-fatty acids usually provide the basis for their capability to cross-link oxidatively. Well-known representatives are linseed, soya, dehydrated castor, wood and fish oils [15].

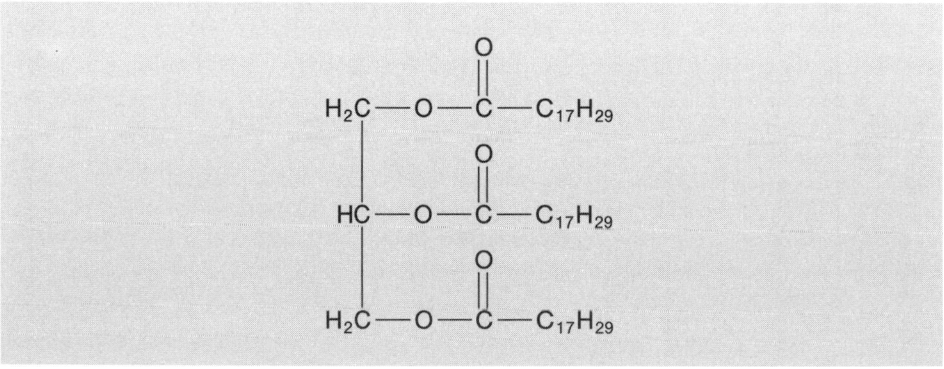

$$\begin{array}{l} H_2C-O-\overset{\overset{\displaystyle O}{\|}}{C}-C_{17}H_{29} \\ \;| \\ HC-O-\overset{\overset{\displaystyle O}{\|}}{C}-C_{17}H_{29} \\ \;| \\ H_2C-O-\overset{\overset{\displaystyle O}{\|}}{C}-C_{17}H_{29} \end{array}$$

In the simplest case, to produce aqueous binders of this basis, the emulsification can be done in water. This is easily possible with the aid of suitable emulsifiers. After addition of dryers and further additives, coatings materials can be obtained that to be sure, display only modest drying and resistance properties.

To improve their properties, chemical modifications of the oils is undertaken. Maleic and fumaric acids, cyclopentadiene, phenolic resins, rosin, styrene, (meth)acrylic monomers, and a range of further compounds are used [13, 16, 20]. Thermal polymerization of the oils is also used to increase the molecular weight, and therefore improve the stability. These so-called stand oils can be further modified chemically. For greater depth, the publications cited in the literature index for chapter 21 are recommended. Here, only the reaction with maleic anhydride will be dealt with, since this reaction is frequently used with a variety of other binders, e.g. with alkyd resins. The relatively easily controlled reaction of maleic anhydride with the unsaturated fatty acids of the oils leads to the two structures shown below.

In both cases anhydride structures are obtained, that can be opened to free acids or half esters with water, or better, with alcohols. Neutralization of the acid groups with ammonia, amines or alkalis gives the water-soluble binder [43].

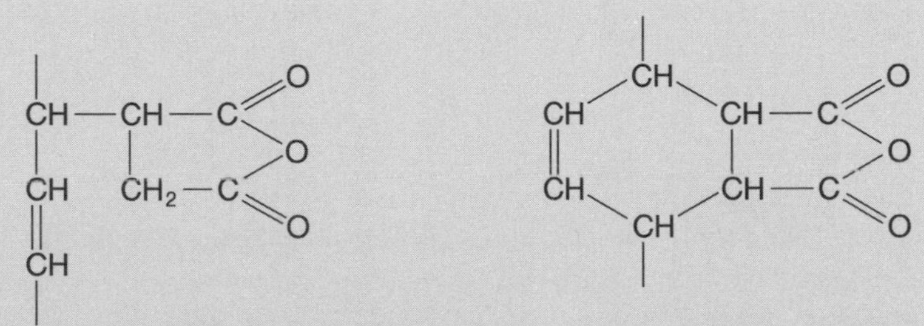

En Reaction with Non-Conjugated Fatty Acid

Diels-Alder Reaction with Conjugated Fatty Acid

Maleic acid-oil adducts, and the above mentioned reaction products, play only a small role as binders for aqueous coatings. The advantages, such as low price and natural raw materials, are countered by many limitations in use. Their significance is thus more in the use as raw material for more demanding oil-modified binders (alkyd resins, modified rosin and phenolic resins). To be sure, there is a market significance for wood impregnation and painting in the United States where the relatively large exterior and interior surfaces of the typical wood houses demands cost-effective paints. In Germany and Europe, there is a certain renaissance through oil-based, aqueous "bio-products".

7.8 Modified Polybutadienes

Polybutadiene can be polymerized by various mechanisms to polymers of differing structures. For the coatings area, products with a high 1,4-cis proportion and molecular weights of 1500–3000 g/mol have won approval (polybutadiene oils). The high double bond content offers multifaceted opportunities for chemical modification. Singled out here are the reaction with maleic anhydride by the en mechanism (see chapter 7.7), the copolymerization with styrene and further monomers.

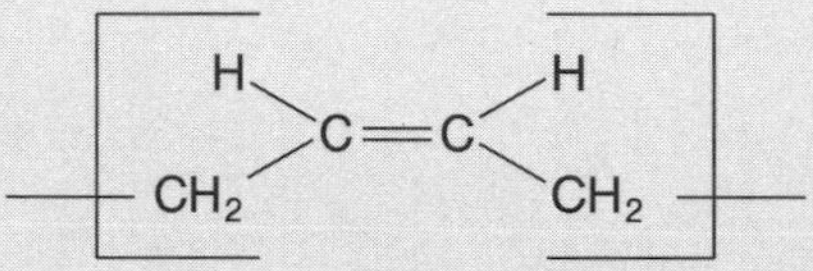

$$\left[-CH_2-CH=CH-CH_2- \right]$$

1,4-cis-Polybutadiene

The acid modified polybutadienes are usually transferred to the aqueous phase in neutralized form (amines, ammonia). "Solutions" as well as emulsions result and corresponding intermediate stages. In addition to these anionic resins, there are cationic polymers, which can be obtained e.g. via. epoxidation of the polybutadiene with subsequent reaction with amines. These resins can be transferred to water by neutralization with acids (e.g. acetic acid) (see chapter 7.4).

$$-C(O)C- + HNR_2 \longrightarrow -C(NR_2)-C(OH)-$$

The chief application for the modified polybutadienes in the area of water-borne coatings is for electrodip priming in automobile manufacturing and in the metal processing industry. Although cataphoretic dip priming based on epoxy resins is today used for practically all automotive chassis, numerous other metal objects from the industrial area, above all from the automotive supply industry, are protected against corrosion with anaphoretic dip paints based on polybutadiene. In addition to this application in the oven-drying sector, aqueous polybutadienes are exceptionally well suited for air-drying corrosion protection priming. Their capability to cross-link oxidatively, paired with saponification stability and good adhesion to nonferrous metals, make them ideal partners for dispersion finish coats. The primers are easy to apply by brushing, usable at temperatures near freezing, and definitely suited for steels cleaned of rust by hand [43]. Depending on whether water-soluble or emulsified resins are used, solvent contents below 10 or 5% can be obtained. For example finish coats of this basis are possible as one-coat paints, but should only be used in interior areas. Otherwise, severe chalking and loss of gloss is to be anticipated due to the high double bond content.

7.9 Phenolic Resins

Phenolic resins are produced by condensation of phenols with formaldehyde and further substances [44]. Alkylphenols permit the production of "oil compatible" types. Although practically all phenolic resins tend to yellow more or less severely, on the other hand, they possess a good resistance to corrosion effects, heat and chemicals. Low-molecular weight methylol compounds of phenol (resoles) are by nature water soluble. To be sure, their storage stability is often limited, so that definite processing conditions must be maintained or less reactive alkylphenols must be used in resin production.

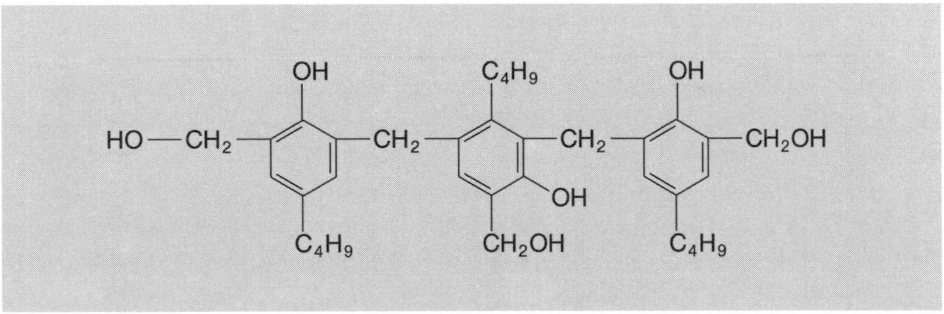

A further possibility for getting water-soluble resins consists in the use of acidic group containing phenols, as e.g. salicylic acid or "diphenolic acid" [13]. All phenolic resin types must be combined with plasticizer resins of another chemical basis. As a rule, the coatings are baked at over 160°C, to obtain the necessary cross-linking reactions. Often the phenolic resins are not used separately in coatings production, but they are used for modifying other coatings resins in their synthesis. The already mentioned good resistance properties of phenolic resin modified binders or paints lead to their use in the most varied baking primers. Darker one-coat paints with high oil and gasoline resistance are also known.

7.10 Poly(meth)acrylates

Polyacrylates and polymethacrylates exist in large variety. There are homopolymers as well as copolymers. A broad range of (meth)acrylic acid esters may be copolymerized with acrylic acid, methacrylic acid and other acrylic derivatives. In addition, comonomers of different types find application. Of particular significance is styrene, leading to the common distinction between pure acrylates and styrenated acrylates. The large number of available monomers makes possible the introduction of differing functional groups in the polymer.

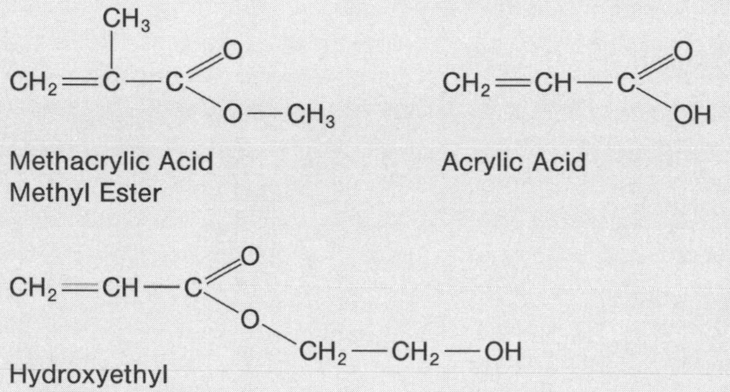

$CH_2{=}C(CH_3){-}C({=}O){-}O{-}CH_3$

Methacrylic Acid Methyl Ester

$CH_2{=}CH{-}C({=}O){-}OH$

Acrylic Acid

$CH_2{=}CH{-}C({=}O){-}O{-}CH_2{-}CH_2{-}OH$

Hydroxyethyl Acrylate

In addition to ester groups hydroxyl, acid, amino, amide and glycidyl groups are common. The nature of the modification determines whether a pure thermoplastic – and therefore exclusively physically drying – or a cross-linkable poly(meth)acrylate is produced [45, 46, 47, 27].

The polymerization process establishes to a large extent how water solubility is achieved. Bulk polymerization, or preferably polymerization in water-soluble solvents, is used to obtain acidic resins. These are of relatively low molecular weight. They become water soluble by neutralization with ammonia, amines or alkalis. Emulsion polymerization is carried out using emulsifiers and protective colloids in water. The so-called poly(meth)acrylate dispersions or emulsions are obtained. These high molecular weight products are water reducible without neutralization [17, 48, 49]. Recently, other production processes have been commercialized. There are now dispersions of solution polymers which are dispersed in water with subsequent distillation of the solvent. These "secondary dispersions" are distinguished by the complete absence of emulsifiers.

The flexibility of their chemistry and means of production has made the poly(meth)acrylates one of the most important classes of binders for water-based paints. Their particular advantage is their hydrolytic stability compared with condensation resins. Physically-drying dispersions are traditionally used as facade paints and other mineral undercoats. Air-drying dispersions and acidic resins are playing an increasingly significant role, often in combination with alkyd resins. Modern dispersions are partially cross-linkable, e.g. with aziridine derivatives [26]. Methacrylate systems are also of increasing importance in industrial baking enamels. Dispersion coatings are enjoying strong annual growth in many countries. The automotive industry is currently preparing for the transition to water-based acrylate clear resins for metallic coatings. Aqueous base coats have already become state of the art practically everywhere.

7.11 Polyurethanes and Polyisocyanates

In the solvent-containing coatings sector, excellent surfaces are obtained with the so-called polyurethane paints. The reaction system usually consists of a combination of a polyisocyanate and a hydroxyl group containing polymer (acrylate, alkyd or polyester resin). The following reaction proceeds thereby:

$$R'{-}N{=}C{=}O + HO{-}R'' \longrightarrow R'{-}NH{-}C(=O){-}O{-}R''$$

Since the isocyanate also reacts with water, such coatings materials are not directly transferable to the aqueous phase.

One never-the-less proceeds in two ways to utilize the properties of the polyurethanes. Thus, one uses the urethane formation during the synthesis of the coatings resin and subsequently transfers this to the water [20]. The so-called polyurethane dispersions result in this fashion. Here it is usually a case of acidic or neutral resins, which are dispersed/emulsified into water directly, or over a solvent step, with neutralization [50,51,52]. Their application takes place alone or in combination with further dispersions in primers, base coats, one-coat paints and special-effects paints. Also a forte is the coating of plastics [53].

$$R'{-}N{=}C{=}O + H_2O \longrightarrow R'{-}NH{-}C(=O){-}OH$$

$$R'{-}NH{-}C(=O){-}OH \longrightarrow R'{-}NH_2 + CO_2$$

$$R'{-}NH_2 + R'{-}N{=}C{=}O \longrightarrow R'{-}NH{-}C(=O){-}NH{-}R'$$

The urethane-acrylate dispersion represents a further development of the technology of polyurethane dispersions. Here, the polymerization of the acrylic monomers is conducted in the presence of polyurethanes. While the urethane portion largely determines the resulting properties, the incorporation of the acrylates contributes a not insignificant materials cost reduction [54]. An extensive presentation of the production and use of these binders, which are constantly increasing in significance, was written by Arnoldus [55].

The second way to utilize the polyurethanes in water-borne coatings is still more complex, since one tries to apply the two-component technique known from solvent coatings. The reaction of the isocyanate with water, with formation of carbon dioxide and N-substituted polyureas, presented above must be suppressed or at least minimized. This is achieved by highly carboxylated fine-particle polyol dispersions to which sufficient low-viscosity polyisocyanates are added [56, 57]. The resulting two-component paint contains a high proportion of urethane groups after hardening and possesses a properties range similar to the well-known solvent systems. Whether these coatings will penetrate the market will be determined by their applications behavior and costs.

7.12 Polyvinylesters and their Copolymers

Vinyl esters are usually used in aqueous coatings in the form of copolymer dispersions. Typical vinyl esters are for example, vinyl acetate, vinyl propionate, vinyl laurate, and vinyl versatate. Acrylic, maleic and fumaric acid esters, ethylene and other copolymers are used. By selecting the monomers and the polymerization conditions, one can obtain a broad range of property combinations [16]. Since the polyvinyl ester dispersions have been on the market since about 1935, there exists today a broad line of quite cost-effective products [58]. The traditional main use is in construction, where these polymers are used in interior and exterior coatings for plaster, concrete, dry wall, etc, but also in plaster and joint compounds. In addition these products are used as dispersion paints for wood, plastics, iron, and non-ferrous metals as well as many additional applications.

7.13 Silicon-Based Binders

Polysiloxanes are produced by condensation of various silanes. The methylsilicon resins represent the best-known resin type. Their special significance lies in the high water-repellant properties with good water vapor permeability and extreme longevity. Applications are primarily in construction, partially in combination with dispersions of other binders [59].

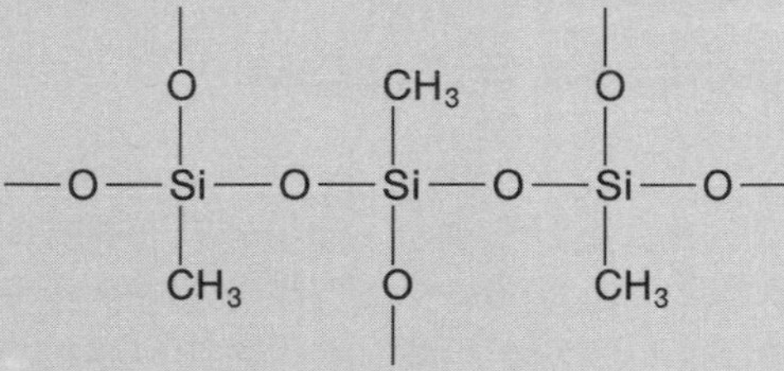

Methylsilicon Resin Structure Extract

They are used to impregnate the most diverse construction materials. Beyond that, they are increasingly found in modern facade coatings [60]. While the silicon resin emulsions or dispersions used earlier still contained solvents, today the so-called micro-emulsions are practically free of volatile components.

7.14 Silicates

From the class of silicates, practically only water glass is used as a binder [16]. Its significance lies with facade coatings. The "pure" silicate samples can be modified with a maximum of 5% synthetic dispersion. Moreover, reference is made to the original literature [61].

7.15 Styrene-Butadiene Dispersions

Copolymerizates of the "hard" monomer styrene and the "flexible" butadiene are known in dispersion form. In addition to the cost-effective monomers, attention is called to the facile control of hardness and flexibility properties. One peculiarity is also the capability to cross-link oxidatively with the addition of dryers. This is based on the butadiene-derived double bond remaining in the polymer molecule, which to be sure, is also the cause of an increased chalking for corresponding paints with exterior applications.

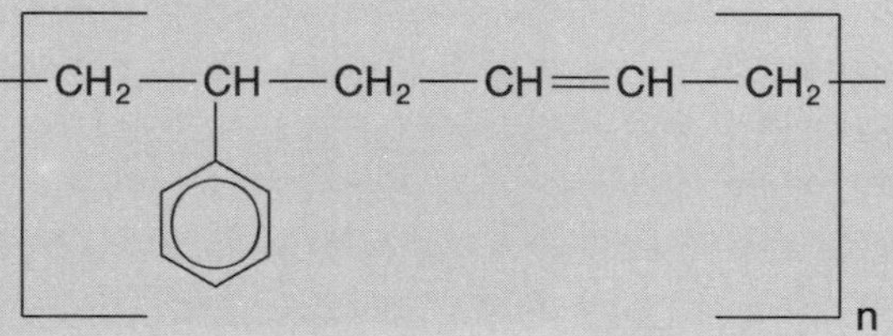

Styrene-Butadiene Copolymer

On the one hand, SB dispersions are used in construction for the modification of cement mortars. Corrosion protection is a further domain of the SB dispersions. Here, dryers are often added to accelerate the oxidative drying of the film and to obtain a low water absorption. The SB dispersions are superior to most styrene-acrylates and pure acrylates when it comes to corrosion protection behavior. Uses are above all in primers, automotive stone chip resistant coatings, and in coatings for trucks and containers [62, 63]. As a rule, a complete paint build up is carried out with a primer and inner-coat based on SB and a finish coat based on styrene-acrylate or pure acrylate. The SB dispersions can of course be combined with other aqueous binders during paint production.

7.16 Additional Raw Materials for Aqueous Binders

Asphalt and bitumen emulsions have long been on the market. They are used above all in construction for coating tanks, roofs and paved surfaces.

Halogen-containing copolymers have a certain significance. Thus dispersions based on vinyl chloride or vinylidine chloride, and frequently further comonomers, are distinguished by non-flammability, good water resistance and low water absorption. Fluorine-containing polymerizates are characterized especially by high chemical and heat resistance and low elongation on heating.

Ethylene is usually copolymerized with vinyl esters (see chapter 7.12.) to valuable products. To be sure, water-soluble products can also be obtained by copolymerization with maleic anhydride.

Similar to waxes, hydrocarbon resins can be dispersed in water with the aid of emulsifying agents. Because of their very good hydrolytic stability and high hydrophobic character, they are suited for impregnation.

8 Additives in Water-Borne Coatings

8.1 Neutralization Agents

The most important method for converting classical coatings resins into water-reducible form is salt formation. With this, the carboxyl groups present in the anionic resin are neutralized with ammonia, amines or amine derivatives. The acid number of such resins is normally higher than that of the usual systems. The influence of the neutralization agent used is important for the stability of the resulting water-soluble resin and its technical applications properties [64]. To be sure, salt formation alone is usually insufficient to attain a stable resin solution that is easy to manipulate. So-called auxiliary solvents are also required (see chapter 8.2. and 10). Because of the relatively high volatile organic substances content of such resins, one is compelled to reduce these additives and thereby the pollution of the environment. By using external emulsifiers, or resins that have emulsifier entities integrated into the molecular skeleton, one arrives at colloidal solutions (see chapter 6.3.). To be sure, here too, neutralization agents and auxiliary solvents cannot be avoided entirely. Due to a number of technical applications properties, as e.g. pigment wetting, drying, gloss and corrosion protection, these systems are in part, not viewed as favorably as the aqueous resin solutions. As is often the case, the user must search out the most favorable compromise between environmental protection and technical quality.

The background for the great significance that the correct selection of the neutralization agent has, is the viscosity anomaly of water-reducible resins (see chapter 10). The neutralization agent has a significant effect on this. The neutralization should be as complete as possible. An insufficient dose results in a greater viscosity peak because of the slight polarity, and thus solubility, of the resin molecules. An overdose has no direct effect on the viscosity peak. The overdose should not be so great as to accelerate the hydrolysis of the binder. The base strength of the amine used is obviously unimportant. Not only the neutralizing effect and salt formation of the amine, but also its effect as a solubility promoter play an important role. The amine influences the storage stability (pH value, viscosity). In general, adjusting the pH value to 8.2 to 8.5 is the most favorable in the case of a solution. For hybrid systems values of pH 7 are also recommended.

As already mentioned, amines or their derivatives are used as neutralization components. Triethylamine (TEA) is frequently used. Ammonia, to be sure, is indicated as a less toxic substitute product, but is also not totally without concern (water pollution, poison class) and, from a purely technical view, has the disadvantage of hardly possessing auxiliary solvent properties. Good results are obtained with mixtures of aminoalcohols (e.g. dimethylethanolamine DMEA and dimethylaminomethylpropanol DMAMP) and ammonia or triethylamine.

Table 7 Properties of Neutralization Agents

Amine	Solubility Parameter δ $(J/cm^3)^{1/2}$	Flash point °C	Azeotrope with water °C	% Amine	Density at 20°C g/cm^3	pK_a-Value at 20°C	pH-Value[1)]
Ammonia		–	–	–	0.8920	9.24	10.0
Morpholine	10.72	39	no	–	1.0020	8.43	9.7
DMEA	9.78	46	91	5	0.8879	9.31	10.1
AMP	10.91	68	no	–	0.94	9.82	10.3
DEEA	9.40	54	99	26	0.8851	9.87	10.3
DMAMP	9.05	57	98	26	0.90	10.2	10.5
TEA	7.46	–7	76	90	0.7290	10.8	10.6
Water	23.00	–	–	–	1.000	7	7

[1)] in 0.005n aqueous solution

DMEA = Dimethylethanolamine,
AMP = Aminomethylpropanol,
DEEA = Diethylethanolamine,
DMAMP = Dimethylaminomethylpropanol,
TEA = Triethylamine

In addition, DMEA has proved to be a good auxiliary solvent and is favored for baking and forced drying systems because of its slow drying [65, 66, 67].

The amines used to neutralize acidic binders differ in their physical and chemical properties (see table 7).

Woods [66, 67] has formulated amine activity factors that indicate the relative amine quantities that are necessary, in switching from one product to another, to achieve a binder pH of 8.5.

Table 8 Amine Activity Factors

	AMP	TEA	DMAMP	DMEA	DEEA
AMP	1.00	0.90	0.79	0.77	0.76
TEA	1.11	1.00	0.88	0.85	0.84
DMAMP	1.26	1.14	1.00	0.97	0.96
DMEA	1.03	1.18	1.02	1.00	0.99
DEEA	1.31	1.19	1.04	1.01	1.00

In general, it should be mentioned at this point, that water-reducible resins, because of their content of amines and solvents that are not innoxious, must be handled with the same circumspection as is customary for classical coatings resins. This relates e.g. to inhalation of spray mists (respirator) and resorption through the skin (gloves) (see chapter 17.1.). To be sure, because of the lower doses, maintaining TLV values is significantly easier and the substance safety criteria under the European Community regulations are usually less critical than with the solvent systems. Also the problem of flammability is significantly reduced due to the dominance of the solvent water.

8.2 Auxiliary Solvents (Cosolvents)

As already mentioned in chapter 8.1. about neutralization agents for aqueous resins, one can not get by with these resins without the addition of solvents (see also chapter 10). The task of the auxiliary- or co-solvents (also known as coupling agents) generally, is to make miscible normally immiscible systems of organic resins and water. Only with their help is the binder producer or formulator in a position to produce stable products with suitable technical applications properties. The water solubility that is attained by salt formation of the carboxyl groups if frequently insufficient to do this. In general, one must realize that with this problem it is a case of a very complex interplay of resin, neutralization agent, degree of neutralization and co-solvent [42, 68]. (see chapter 10). The determining criteria for the selection of auxiliary solvents are the storage stability (viscosity and pH value development) and the technical applications behavior (viscosity, rheology, wetting, gloss, etc.) [69, 70].

Aqueous solutions and resin emulsions behave differently with respect to viscosity, and a varying influence of the auxiliary solvents must be ascertained. Resin solutions display a viscosity anomaly in the form of the water peak that appears upon thinning with water (see chapter 6.1 and 10). The more auxiliary solvent that is used, the less this anomaly results, and finally is completely eliminated. The emulsions mentioned display no anomalous behavior: their viscosity decreases upon addition of water, analogous to the behavior of the usual polymer dispersions. An addition of auxiliary solvents can even lead to a viscosity increase due to swelling of the emulsion droplets. In the following therefore, only resin solutions will be considered.

The reduction in the viscosity anomaly by solvents is caused by the interaction between resin and solvent, whereby the formation of association between the resin molecules is constrained. Not only the quantity, but also the type of solvent is deciding. Favored for use are n-butanol, 2-butanol, and in general, butyl glycol is especially effectively [42, 71, 72]. Good effectiveness is also displayed by many ethers of propylene glycol, such as isopropoxypropanol and n-propoxypropanol (see chapter 10 and table 9) [73].

Solvents also have an important influence on the processing properties of a water-borne coating. Thus some auxiliary solvents display a temperature-dependent solubility for resins, which results in the fact that a different co-solvent may be optimal with air drying than with forced drying or baking. A well-known phenomenon is the sagging of a coatings material in the drying process which is attributable to a decrease in viscosity of the binder upon vaporization of the solvent. Such behavior occurs when the dilution diagram (viscosity/solids content)of the corresponding resin-solvent combination displays a maximum.

Table 9 The Most Important Auxiliary Solvents (Cosolvents) for Water-Borne Coatings

Alcohols	Ethylene glycol ethers	Propylene glycol ethers
Ethanol	Ethyl glycol	1-Methoxy-2-propanol
Propanol	Isopropyl glycol	1-Ethoxy-2-propanol
Isopropyl alcohol	Propyl glycol	1-Isopropoxy-2-propanol
Butanol	Butyl glycol	1-Propoxy-2-propanol
Isobutanol	Isobutyl glycol	1-Butoxy-2-propanol
sec-Butanol	Ethyl diglycol	
tert-Butanol	Butyl diglycol	

The dilution diagram can thus provide an indication of the behavior to be expected upon drying. One must also be aware of the control of the rate of the drying process with auxiliary solvents (see also chapter 14) [74]. Solvents thus improve the rheological properties of a water-borne coating (flow, sagging, edge wetting) [75]. Butyl glycol is often used as a suitable agent to reduce edge evaporation, which is especially pronounced with aqueous systems and which reduces the corrosion resistance by lowering the film strength on especially exposed edges [76). To be noted is the fact that auxiliary solvents have drawbacks from an environmental point of view, since they threaten water, even if usually only slightly. With respect to flash point, there is in general a significant advantage over classical solvent systems. Thus, for example, recipes using butanol or butyl glycol with a total solvent content of less than 15% are not subject to labeling requirements under the EC regulations. Since a minimum solvent content should always be targeted, flammability should play no role (see chapter 17.2)

8.3 Film Forming Agents

Due to the special nature of polymer emulsions, the film forming process deviates substantially from classical coatings resins dissolved in solvent and also in water [77]. Since emulsions consist of polymer spheres and not discrete molecules, drying must involve fusion of these relatively large particles. This process can be divided into several steps [78,79]: In the beginning, the polymer particles increasingly approach each other, since the total volume decreases because of the evaporation of water. To achieve direct contact, the electrostatic repulsion forces between the spheres, which originally provided for the stability of the dispersion, must be overcome. After mutual contact, a sufficient deformation of the particles for arrangement in a film and a subsequent fusion is never-the-less necessary for a true film formation. This only occurs when the capillary forces and the surface tension forces are larger than the deformation resistance, i.e. the hardness of the spheres. The latter can be well described by the glass transition temperature T_G of the polymer, which in turn is closely related to the minimum film-forming temperature MFFT, a specific value for every dispersion.

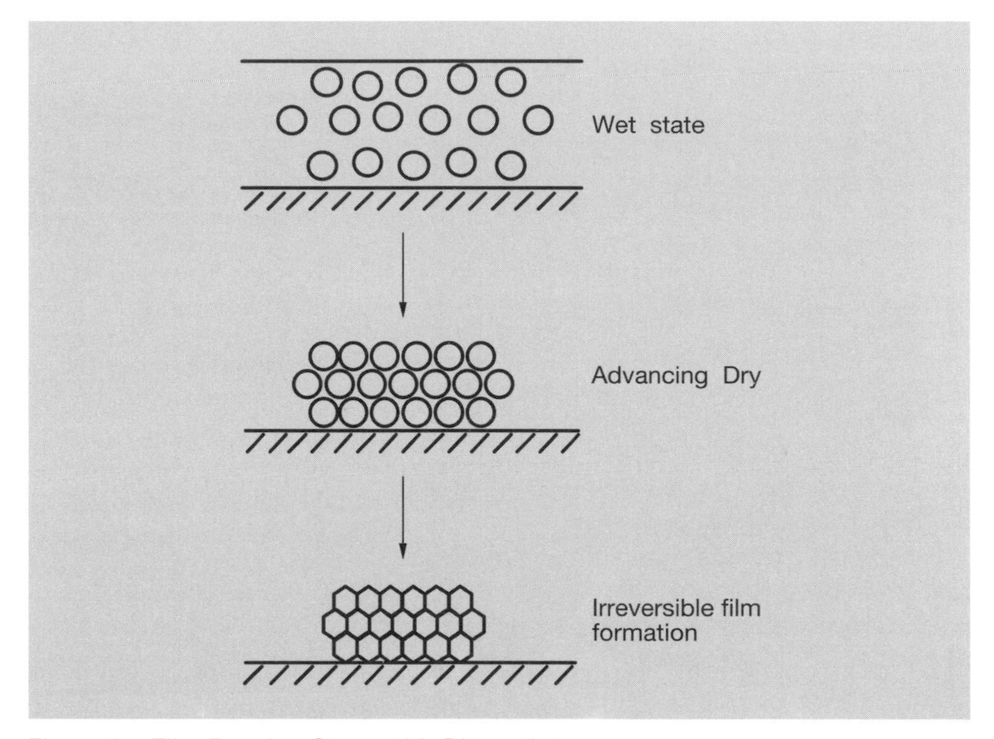

Figure 3 Film Forming Steps with Dispersion

It is inviolate, that the dispersion particles can only form a film above the T_G of the polymer or more precisely above the MFFT. Thus, one is compelled to control the processing temperature T_P of the polymer, by a corresponding selection of the monomers (internal plasticizer) or the addition of external plasticizers, so that :

$$T_P \geqq \text{MFFT}$$

Because of this handicap, one is constrained to, with regard to the final hardness, a glass temperature and degree of hardness that is not adequate for many applications. This theoretical principle is circumvented by the addition of so-called film-forming agents and coalescing agents [13]. Such substances, also designated temporary plasticizers, provide for a softening of the polymer particles, but due to their volatility leave the film after film forming. In this way, dispersions of a hard co-polymerizate with

$$T_G > T_P$$

can be made to form films at room temperature and subsequently the desired hard coating can be obtained.

For a product to be used as a film-forming agent, it must satisfy definite requirements [80,81]. A prerequisite is a sufficient chemical stability of the temporary plasticizer so that for example, saponification is precluded in the usually alkaline recipes. Today, in addition, low odor, above all for interior applications, is demanded. Further a film-forming agent must be compatible with the usually sensible polymer dispersions, so that an instability does not develop. These can manifest themselves as a severe increase in viscosity, or in the worst case, coagulation of the binder. For these reasons, manipulation should be undertaken carefully, to avoid local shock reactions (coagulation). In general, an aging time is required after producing the final product, so that the distribution of the film-forming agent can reach a state of equilibrium in the paint. The temporary plasticizer should have as great a plasticizer effect on the corresponding polymer as possible, in order to be able to keep the dosage small. With respect to evaporative behavior, the ideal is to have the film-forming agent remain in the film during film formation and to act as plasticizer, but to have it evaporate as soon as possible after film-forming is complete. If a film-forming agent remains in the film too long, this will lead to tackiness, and thus to a tendency to block and soil [82]. Since, as can be anticipated, the ideal film-forming agent does not exist, the optimum compromise, perhaps with a combination of products, must be sought.

The so-called three-phase model describes well the behavior of film-forming agents [83]. This model proceeds from the fact that a substance used as a film-forming agent distributes itself in a definite relationship among the three phases of water, particle nucleus and particle shell. Residence in the shell of the polymer particle is decisive for an optimum effect, since this is where film formation takes place. If the film-forming agent is very water-soluble and thus is primarily localized in the water phase, the plasticizer effect is naturally slight. If it is located preferentially in the nucleus of the polymer particles, its influence on film-formation is again modest. In addition, in this case, water solubility is usually very slight, so that the manipulation which must proceed in the aqueous phase is problematical: The product wanders only slowly from the water into the polymer particles and the film-forming agent in droplet form can lead to turbidity in the film with insufficient aging. An alternative here is the addition of such a temporary plasticizer during polymerization (dosage dissolved in the monomer).

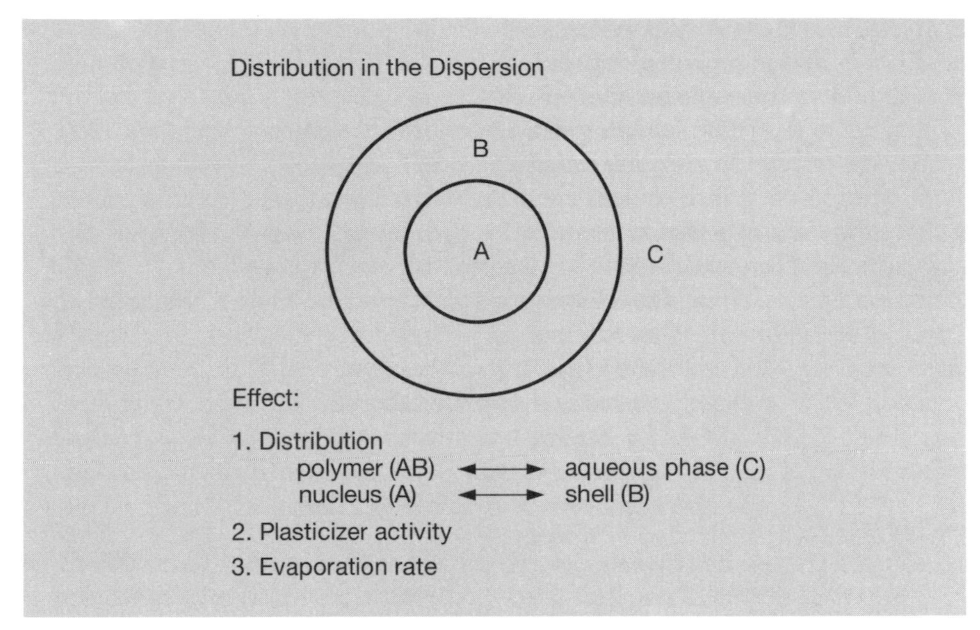

Figure 4 Three-Phase Model to Display the Activity of Film-Forming Agents

To be able to judge the suitability of a product as a film-forming agent, one must have indications of its distribution in the dispersion, its plasticizer effect [84] on the polymer and its rate of evaporation [85]. The evaluation of these indications will certainly differ from dispersion to dispersion, but also from application to application, so that in sum, the universal film-forming agent does not exist. Not lastly to be explained as a result of this fact, is that there is today a multitude of products that are recommended as coalescing agents. Table 10 gives an overview of the groups of film-forming agents, their distribution in the dispersion (figure 4) as well as their effect on film formation and further technical applications properties, as e.g. the scrub resistance with emulsion paints. It should be clear, that this is only a cursory classification. Never-the-less important indications are present. Table 10 makes clear that aromatic and aliphatic hydrocarbons, which are primarily localized in the particle nucleus, have only a moderate film-forming effect, while the esters and ketones, which are also present in the shell, show the highest effectiveness. Ether alcohols, while not as effective as film-forming agents, are never-the-less in a position to control the open time of a water-borne coating. For products with good water solubility, there is the danger of separation in the case of an absorbent substrates, and so the dissolved components cannot provide any type of contribution to film formation. With non-absorbent substrates such as metal, the behavior is different: In spite of good water solubility, film-formation is supported here. There are further technical applications properties to be named that can be influenced by highly water-soluble products: Drying behavior, rheology, freeze-thaw stability, inter. alia. Such substances should be considered as boundary cases between film-forming agents and co-solvents.

These indications should have made clear, that there are no firm rules, but only reference points for the use of film-forming agents. It is the task of the formulator to select the product that is optimum for his binder and his recipe as well as for the particular application.

Table 10 Types and Effects of Film-Forming Agents (Distribution see figure 4)

Type	Hydrocarbon	Ester ketone	Ether alcohol	Glycols, hydrophilic ether alcohols
Distri-bution	A	AB	ABC	C
Film formation	Moderate	Very good	Good	Slight
Effects	Difficult manipulation, modest gloss development	Good scrub resistance, tackiness, limited open time	Good open time, tackiness, worse dew resistance	Good open time, good freeze stability, worse dew resistance
Examples	Reference gasoline isoparaffin, cycloaliphatics	Texanol; adipic, phthalic, and benzoic acid esters of propane diol	Propylene glycol ether, butyl glycol, butyl diglycol	Methoxy-propanol; ethylene, propylene and diethylene glycol

8.4 Preservatives

Doubtless, the increasing trend, "Down with solvent-containing coatings systems and up with water-reducible products!" is contributing to a reduction in environmental pollution and in the risk to the processor. The use of preservatives in water-borne coatings entails the introduction of not necessarily innocuous chemicals in the coating. The sensitivity of all concerned with this difficulty is expressed by the constant search for new less-hazardous, but never-the-less effective, substances. It is however accepted, that such additions are unavoidable under prevailing conditions. Water-reducible systems are constantly threatened by attack of micro-organisms, but must meet the usual storage period requirements. Paint residues in containers must also remain resistant [13].

Making things more difficult, is the fact that water-reducible binders contain ever smaller amounts of contaminants such as residual monomers that appear to have biostatic activity. Thus raw materials suppliers [86], as well as the paint and coatings industry, must strive to preserve their products by the judicious use of preservatives.

Typical failures of contaminated products are above all, discoloration, odor and gas evolution. Attack can also be manifested however, through viscosity decrease, sedimentation, coagulation effects, changed rheology and pH changes. To reduce contamination, it should be noted that sources of contamination are widespread. Thus micro-organisms can be present in water and cellulose ester thickeners as well as in pigments and fillers. So-called stock solutions (e.g. thickeners or color pastes), which are frequently stored for extended periods are suspect. The processing facilities [87] themselves are critical, as for example blind angles in pipelines or crusts in mixing kettles or storage tanks. Finally, air also contains bacteria which can form on moist walls for example. In general, surveillance of all raw materials and regular cleaning of the facility is recommended to maintain contamination as low as possible from the start. Now as in the past, dilute formaldehyde/hypochlorite solutions have an optimum effect for such purification measures.

What requirements should be made of a preservative? Since the threat is not posed by only a few definite bacteria, the product to be used should have as broad a spectrum of effectiveness as possible. Further, a quick and lasting effectiveness is desired. In general, it must be noted, that preservatives are sensitive to boundary conditions such as pH, temperature, re-dox potential, etc. These constraints should be as slight as possible. From an applications standpoint, it is important that the product be very water soluble and be completely compatible with the binder and the total formulation. Tests, such as investigating the binder film for specks and coagulation for example, as well as checking the storage stability, can give an indication of this. It should be noted that the actual preservative concentration can lie significantly below the amount charged due to degradation or incompatibility. Sufficient preservation may then no longer be guaranteed. It is considered advantageous, if the agent shows a certain activity in the gas volume. As indicated previously, increasing weight is being placed on the environmental compatibility of preservatives, including reduced human toxicity (skin irritation, skin sensitivity, oral and inhalation toxicity), reduced water pollution (fish toxicity, organically-bound chlorine) and a deleterious effect on the activity of water purification facilities. For a few applications, care must be taken that the biocide corresponds with FDA guidelines.

Formaldehyde, valued now as in the past because of its effectiveness, is under pressure. Phenolic compounds are also hardly used any longer. The current spectrum of products that are used includes above all, reaction products of alcohols, amides and amines with formaldehyde, isothiazolinones, benzisothiaazolinones and chloracetamide.

With regard to the amounts used, it is important to note that an overdose is ecologically and economically senseless, but that too small a dose can be dangerous. If the concentration of the preservative lies below the minimal inhibiting concentration, the effect will not suffice to contain the multiplying of the bacteria that are always present. If the bacterial count begins to rise exponentially, a significantly higher amount of preservative must be used to kill off the bacteria. It is advantageous to have a second preservative with a different mode of action available because of the danger of building resistance.

The various microbiological test methods will not be dealt with here [88]. Is will only be mentioned that easily used test sticks [89] are available in the market to determine bacterial counts. By this means, raw materials as well as end products can be monitored without great effort.

If coatings materials are to be protected against fungi and algae, adjuncts can again be used. At the present, similar products are used as preservation in the can, but in higher concentration [13, 90]. The adjuncts must have a certain water solubility, so that they are remain effective when soil layers, which provide nutritious surface, build on the coating for example. This solubility results in the washing out, after some time, of the fungicides and algicide and the loss of their activity. They thus do not convey lasting protection. Here too, the material must be compatible with the recipe. Care must be taken, however that as few materials as possible be introduced that can serve as nutrients. Manufacturers provide recommendations according to the application area (interior, exterior, food areas).

8.5 Thickeners

The concept of thickener imprecisely describes the functions of the auxiliaries to be described here. Three key effects can be differentiated:

- thickening effect, i.e. the pure viscosity increase
- rheology control
- reducing settling of pigments and fillers.

Naturally, these effects are not clearly separable, since for example, each classical thickener also has an effect on rheology and reduces the tendency for settling. But, they point out the general direction. Thickeners are used with typical dispersion paints to establish a favorable processing viscosity, since dispersions as binders usually do not have adequate viscosity. The addition of a rheological agent to emulsion paints is necessary to obtain flow properties that approximate those of classical paints as closely as possible. Aqueous resin solutions usually have sufficient viscosity and the desired rheology for use in water-borne coatings. To be sure, an anti-settling agent is needed to prevent the formation of a hard sediment. In addition to accomplishing the tasks mentioned, thickeners must not adversely affect the technical applications properties of the coatings and paints such as water uptake, corrosion protection, soiling, gloss, foaming, scrub resistance, storage stability (attack of microorganisms, settling) etc.

Before getting into the individual types of thickeners, it must be remembered that the fillers used in paint also have a significant influence on viscosity and rheology (see chapter 9.5). This should always be noted, since fillers are significantly more price-effective and in general have fewer side effects than thickeners. Rheological effects can also be controlled with the addition of small amounts of solvents. Finally it must be noted that resin manufacturers are actively trying to develop binders that convey favorable rheological behavior and thus reduce the need for the addition of additives.

Thickeners will be divided into four groups:

- inorganic
- organometallic
- organic and
- synthetic organic products.

8.5.1 Inorganic Thickeners [91]

The most important representatives of this group are the stratified silicates. Montmorillionite bentonite (Al-silicate) and hectorite (Mg-silicate), as well as attapulgite (Mg-Al-silicate) should be mentioned. Above all, these products display a thickening action that arises by formation of a gel structure. The gel develops through the interaction of differently charged particle surfaces and partially through swelling of the particle (deposition of water in the interfaces). The gel structure is very marked but is reduced by shear forces. Upon discontinuing the shear it rapidly builds up again, so a structural viscosity rather than a thixotropic behavior is portrayed. Mixtures with cellulose ethers or associative thickeners are advantageous for use in paints and coatings. While the latter give good flow properties to the system, the stratified silicates provide a certain thickening and reduce the tendency of fillers to settle. As in general with inorganics, the resistance to microorganisms is noteworthy. Slight water uptake is naturally combined with slight water retention. Compared with the cellulose ethers (high water uptake, good water retention), the stratified silicates have a better influence on the scrubability of dispersion paints.

Additional representatives of inorganic thickeners are the fumed silicas, which are produced by hydrolysis of silicon tetrachloride and are formed as extremely fine amorphous powders. The primary particles measure in the nanometer realm. The agglomerates formed from them lie between 0.1 and 1 µm. The present products differ in their specific surface and special surface treatment (hydrophilic and hydrophobic products). The thickening effect arises through formation of a network which develops via. hydrogen bonding between the Si-OH groups present on the surface. Systems thickened in this manner display thixotropic behavior. Structure formation depends on the polarity of the medium. The effect is greatest in nonpolar systems, while it is above all reduced in water by the interaction with its hydrogen atoms. Fumed silicas are used when a high coating thickness is needed or when a large thickening is desired. Additional noteworthy properties are the slight water take-up and slight water retention as well as resistance to microorganisms. Incorporation of fumed silicas is not without difficulties because of the low bulk density. Despite the fine particles, there is no danger of silicosis.

8.5.2 Organometallic Thickeners

These are products the do not work alone, but in combination with carboxyl groups present in the polymer, and above all with protective colloids such as cellulose ethers and polyvinyl alcohol. Gel structures result from hydrogen bonding between the polymer particles and the metal complexes [92, 93] (above all titanate and zirconium complexes). Typical emulsion dispersions can not be modified directly. If, as customary with classical emulsion paints, cellulose ethers are used as thickeners, the addition of titanium chelates, for example, will result in definite gel formation. Above all this product group has attained significance for the production of spatter-resistant paints. It is noted that the gel structure reforms slowly after the application of shear forces (e.g. with brushing), so that the paint can flow in the interim and the brush marks can disappear. Such complexes have also proven advantageous in combination with associative thickeners. Paints are obtained that have low sagging, good flow and slight tendency to splatter.

8.5.3 Organic Thickeners

The cellulose ethers, which above all are usually used in classical dispersion paints, are the most significant of these [13, 81]. In these products, anhydroglucose entities form catenated macromolecules with a degree of polymerization between 50 and 1000. The latter determines to a large extent the thickening effect. The higher it is, the higher the viscosity of a cellulose ether solution. The thickening activity is usually not described by the degree of polymerization, but by the increase in viscosity of a 2% solution. In addition to the length of the polymer chain, the type of ether formation and the degree of substitution have a significant effect. Converting the hydroxyl groups present in the glucose molecule to ethers results in carbomethoxy, methyl and hydroxyethyl celluloses as well as mixed products such as methylhydroxyethyl and methylhydroxypropyl celluloses [94]. The most important products are methyl and hydroxyethyl cellulose. The solubility of the diverse nonionic cellulose types can be controlled by the type and degree of substitution. Cold-, warm-, as well as warm- and cold-soluble products are available.

Generally, these are powders that can be stirred into water. The so-called delayed types were developed to minimize lumping. These are initially stirred in at low pH and distributed well. Only subsequently is the pH raised. The cellulose ether dissolves, as a result of the increased cross-linking with glyoxal, and has a thickening effect.

In general, it can be established, that cellulose ethers above all have a thickening effect that with respect to rheology, achieves a structural viscosity rather than thixotropic properties. Again, these effects depend strongly on the degree of polymerization and substitution, the type of ether and naturally, the amounts used. Cellulose ethers can be mixed with associative thickeners, for example, to optimize rheology. It must be noted that cellulose derivatives have high water retention, which is favorable for good processing. Correspondingly however, water uptake is high in the applied film. As a result they are susceptible to microorganisms, so that care should be taken with the doses used. To be sure, cellulose ethers also have a stabilizing effect. Like polyvinyl alcohol, they are conventionally used as protective colloids in synthetic dispersions. In analogous fashion, they also stabilize pigments and fillers in paint recipes.

For the sake of completeness reference is made to additional thickeners that are partially related to the cellulose ethers. Among these for example, are the xanthanes which are produced by fermentation of glucoses, mannoses and glucuronic acid with bacteria [95]. These thickeners convey a pseudoplastic flow behavior to solutions, i.e. a structural viscosity with a flow boundary. They can thus be used where this special behavior is desired, for example in drip-free paints. Alginates are also polysaccharides that are, however, derived from mannuronic and guluronic acid. Their significance today is as slight as that of the guar derivatives (galactose and mannose entities). Starches are used less in classical brushing products than in paper inks, where they have more of a binder character than that of a thickener. As a result of the "bio-wave", casein, as a natural product, is undergoing a certain renaissance. It was previously used in latex paints and provides favorable processing properties through good flow and gloss. The thickening effect is not large, so that correspondingly high doses are required. Since casein is very susceptible to microorganisms, it requires extensive amounts of preservatives. Although natural, its use as an especially environment-friendly product must be viewed with a certain amount of skepticism.

8.5.4 Synthetic Organic Thickeners

The primary representatives of this group are the polyacrylates [96] and polyvinylpyrrolidones as well as the previously mentioned associative thickeners based on polyurethanes [97] or styrene-maleic acid polymerizates [98]. Polyvinyl alcohol also has a thickening effect. It is used less in the form of an additive, but is introduced directly during polymer production (protective colloid) as a stabilizer and thickener. Compared with the cellulose ethers, all of these products have in common a relative resistance against the attack of microorganisms. On the other hand, they display no significant water retention and are correspondingly more water resistant than the cellulose ethers. In general, it has been established that the synthetic thickeners interact more strongly with dispersions and thus must be matched with the binder as to type and quantity.

Included in the overall concept of polyacrylates are polymers that contain acrylic acid, methacrylic acid and acrylamide. Low-molecular weight products are also used to stabilize pigments in aqueous paint products. Only with higher-molecular weight types, does one observe a thickening effect which, similar to cellulose ethers, is dependent on molecular weight. These are systems, which analogous to known aqueous alkyd resins, become water soluble by neutralization of the acidic entities they contain, or they are true, relatively low-solids dispersions of corresponding polymerizates. The products are usually stirred into the recipe to be thickened in a low-viscosity state and subsequently the pH is made alkaline as required for thickening. With respect to rheological properties, the thickened systems generally display non-Newtonian flow behavior. The property range of soluble polymers is strongly dependent on the degree of coiling of the macromolecules and thus also the pH and electrolyte content of the water-borne coating. In this respect, dispersions which act as thickeners display reduced sensitivity and lead to gelatinous thickening with flow boundaries (pseudoplastic behavior). Further, there are special types that are supplied in organic carriers (aliphatics, plasticizers), do not need additions of alkalis and are effective over a relatively broad pH range. In addition, polyacrylates that have been made hydrophobic (hydrophobic alkali-soluble emulsions (HASE)) have been developed [99]. Compared with the usual acrylates, these have better rheological properties and approach polyurethane thickeners at lower cost. Thickeners based on polyvinyl pyrrolidones are also effective in acidic ranges. The end products produced with these display thixotropic flow properties.

The so-called associative thickeners [100] have achieved great significance in recent years. Their effect is less in the area of thickening than in the establishment of definite rheological properties. Cellulose ethers or acrylate thickeners alone do not provide a satisfactory range of properties, especially for water-borne coatings that are naturally compared with classical coatings resin systems. The viscosity is relatively high with modest shear forces; a flow boundary exists in part. However, with increasing shear, the viscosity drops off strongly. In contrast, associative thickeners provide a low rest viscosity and a slower drop-off in viscosity with higher force as well as a more Newtonian behavior. For brushing applications, this means an increased brushing resistance, which is reassuring to the processor, since thicker films are possible. And after brushing, the paint has better flow (no brush marks). At the same time, the tendency to spatter is reduced (e.g. with roll application) and gloss is higher as a result of the smoother surface [101]. Styrene-maleic acid polymerizates and polyurethane products, which are distributed in Europe especially, occasion such behavior [97]. The later are also designated as HEUR (*h*ydrophobically modified poly-(ox*e*thylene) *ur*ethane) [102]. They are constructed of three components: a hydrophilic back bone (e.g. a polyethylene glycol chain), hydrophobic segments and urethane entities. The difference with other thickeners consists of the fact that they do not thicken the aqueous phase by immobilizing the water molecules. Associative thickeners build a skeleton within the painting system, in which the hydrophobic segments can both merge to form micella and can be absorbed on the surface of the dispersion particles. The skeleton thus couples micella and polymer particles and is less rapidly degraded by shear forces than the usual thickeners. This makes clear, that such products must precisely fit the binder being used. Also to be noted is the fact that other additives such as wetting and dispersing agents, as well as auxiliary solvents affect associative thickeners, in that they affect the formation of micella. While PU thickeners can be used exclusively in water-borne coatings, mixtures of cellulose ethers, acrylates or stratified silica with associative thickeners are recommended for classical emulsions.

8.6 Wetting and Dispersing Agents

To optimize the effect of pigments and fillers with respect to tint, gloss, hiding power, durability, storage stability, etc., these substances must be finely dispersed in the coating material [103]. For this, three prerequisites must be fulfilled: The pigment must be well wetted by the medium, the pigment agglomerates, that have been built by assemblage of the primary pigment particles, must be divided and finally, the fine dispersion must be stabilized, e.g. the re-agglomeration must be prevented. Wetting of mineral fillers and pigments, which generally have hydrophilic surfaces and currently are made to have good wetting and dispersing properties by special treatment, usually proceeds in aqueous media without difficulty [104]. Thus, the addition of a wetting agent to reduce the surface tension is not always necessary. Today they are even recommended against due to their tendency to foam. However, with organic pigments or specially treated fillers, the use of wetting agents is sensible. A precise check of the quantities used is necessary to avoid making the total system too water friendly, which would be especially disadvantageous for corrosion-protecting products. Special aminoalcohols, which due to their volatility do not remain in the film, also show good activity as wetting agents. Mechanical division of the agglomerates proceeds through dissolvers, mills or roller mills, whereby it must be observed that only agglomerates, and not aggregates, that is, tightly-bound particles can be smashed: The fineness of the pigments and fillers is thus limited by the mill fineness of the product supplied. The decisive step is then the stabilization of the dispersed particles by the addition of dispersing agents [91, 105]. Here, above all, polyphosphates, borates, polyacrylates and maleic acid-copolymerizates have prevailed. Less effective are the so-called dispersing resins, which display a similar mode of action as the above products, but remain completely in the binder after drying.

The dispersion agents work in the manner that they are absorbed on the pigment surface and encase the dispersed particles [106]. The prevention of a new agglomeration (flocculation), and thus stabilization, results from electrostatic repulsion. Correct dosage of the dispersing agent is deciding for stability. For its determination, one determines the viscosity minimum of a pigment slurry as a function of the dosage of the additive [81, 107]. If less dispersing agent is added, the system is not optimally stabilized. If too much is added, flocculation can arise due to discharging the stabilized envelopes, e.g. falling off of the repulsive forces. It is recommended to always use a dosage just above what has been determined to be ideal to prevent product variation. Mixtures of polyphosphates and polyacrylates have proven themselves with dispersions, whereby the polyphosphate also acts a water softener: Multivalent cations can favor flocculation.

8.7 Defoamers

Foams, according to the definition of colloid chemistry, concern gases that are finely divided in liquids or solids [91]. The first case, e.g. air dispersed in binders or in the paint, stands in the foreground of consideration, since only here is there the possibility of counter-measures. If these measures do not suffice, the second case will be present after drying of the binder: Air is dispersed in the solid, can no longer be removed and impedes the protective function of the paint. Both types of foams lead to problems of differing types and must be avoided. The foam difficulty becomes most visible with filling processes: Due to the increase in volume, the volume-weight relationship of a foaming product no longer holds, so that the containers or tanks are suddenly "too small". Disadvantages can also be ascertained in the applied product, whereby it can be a matter of purely optical failures or grave technical shortcomings.

The formation of foam in coatings has a cause in principle: Air stirred into the liquid medium with mechanical processing, can not leave sufficiently rapidly [13, 91]. This effect usually occurs with pumping in paint manufacturing and also during application [108]. The dispersed air bubbles differ in that at times they may surrounded by a stabilizing film of surface-active substances or they may be "naked". In general, the air bubbles, depending on their size, have the tendency to rise to the surface of the liquid where the "naked" bubbles burst. The stabilized gas bubbles, however, collect there and initially form a so-called spherical foam, which after joining with larger portions of the inclusive surface film, is transformed to a polyhedral foam. This then is the phenomenon that is customarily known as foam. It becomes apparent, why especially dispersions that have certain amounts of free emulsifiers especially tend to foam: The emulsifiers stabilize not only the polymer spheres, but unfortunately also the air bubbles.

For a detailed consideration of the foam phenomenon, it is useful to differentiate between macro and micro foams, wherein the terms "macro" and "micro" refer to the size of the bubbles [108]. As already mentioned, the rate of rising is a function of their size. Thus the above refers primarily to large gas bubbles which rise relatively rapidly and then form the macro-foam. Single large bubbles can lead to cratering in the coating if the coating material is sufficiently rigid upon bursting that it can no longer flow together. Less obvious, but not less problematical, is the behavior of small air bubbles. These are more stable than large bubbles due to their smaller diameters. They diffuse very slowly and are at times not able to penetrate to the surface of the liquid. Thus they remain as a micro-foam in the paint film, which can lead to a concentration underneath the surface. Or they leave so-called pin holes at the surface and in the film, which arise from the fact that the small bubbles, because of their slow rise, do not reach the surface until the coating is already so viscous that the holes no can longer coalesce. A type of chimney is thus formed which reaches to the substrate. This makes clear that this is a matter of a technical shortcoming with significant consequences, whereas the problem with craters that has been mentioned is more an optical one. A trained eye or a suitable analytical method is necessary to recognize micro-foams. Thus pin holes can be easily identified if the coatings material is spread on a glass plate and observed against the light.

There are certain influences which promote the formation of foams or prevent defoaming with the help of defoamers [81, 109]. Thus dispersions are known to tend to foam more strongly with increasing pH and reduced particle size. The pigments and fillers used in paint manufacture contribute air to the system, as a function of their size, and thus promote foam formation. Thus to get better de-gassing during paint manufacture, fine fillers should be used first and finally large fillers. Ideally, if not always practical, is the use of vacuum dissolvers by which foams that arise during processing can be eliminated. The coatings properties, the applications method and the drying procedure also are of influence. At first glance, it is to be expected that highly viscous, high solids coatings will behave relatively unfavorably. Highly viscous air-less coatings are usually more difficult to defoam than are coatings sprayed by air pressure. Because of the longer drying process, which gives the air bubbles a relatively long time to escape, air drying is more favorable in this respect, than are forced drying or baking, where the bubbles are quickly immobilized. Reference should be made here to the tendency of water-borne coatings to form blisters, which occur when the evaporating water cannot continuously leave the film because of the rapid surface drying, but penetrates through the surface explosively.

The auxiliaries used for foaming problems are usually all designated as defoamers. As discussed above it is sensible to distinguish between defoamers and de-gassing agents. The function of the defoamer is to degrade the foam formed on the surface. This proceeds by the defoamer destroying the surface films, which consist above all of surfactants, and which stabilize the foam bubbles, and thus bursting them. The mechanism, which is not clarified in all its details, will not be discussed here [110]. De-gassing agents enable the captured air to reach the surface from the inner coating film. Thus, defoamers are sufficient to combat macro-foams, while a combination of de-gassing agents and defoamers is recommended for micro-foams.

An important prerequisite for a substance to be a defoamer is in general, its insolubility in the medium to be defoamed. Accordingly, it is reasonable that product groups such as fatty acid esters, metal soaps, mineral oils, waxes, silicone oils and siloxanes, potentially in combination with emulsifiers and hydrophobic fumed silicas, are used in defoamers which are frequently of complicated composition [91]. Additional important requirements are good compatibility with the coatings material and good effectiveness after extended storage. Here the user must be clear that in the selection of the appropriate defoamer is frequently a matter of choice between two extremes: a product of optimal effectiveness, which due to its high incompatibility tends to form "fish eyes", or a stable more compatible type that has less activity. To be noted also, is that excessive dispersing during paint production can reduce the effectiveness of the defoamer, while poor processing can lead to surface disturbances (defoamer droplets on the film surface: fish eyes) after a certain shelf time. Here, a step-wise addition of defoamer before dispersing, and also in the painting phase at the end, has proven out. Accordingly, the defoamer composition and theoretical considerations (there exist recommendations for special defoamer types for dispersion paints and water-borne coatings) are of less significance to the practitioner than the precise evaluation of its behavior in the relevant formulation. Test methods to do this have been developed by the producers of defoamers [111]. Thus the effectiveness of a defoamer in the production and filling-off of paints can be simulated by a stirring test. The roll test serves to control the behavior during processing. Here wet and dry films are checked for craters and blisters and also for the degree of gloss. Also the so-called rub-out test for color acceptance can be carried out as a function of the defoamer. All tests should be repeated after a certain shelf time. The wetting and dispersing agents used in aqueous systems have a not insignificant effect. Here there are special products that act simultaneously as wetting agents and defoamers. It is noted that isoparaffins [112], which are used to improve substrate wetting, to control the open time and to some extent as film-forming agents, have a defoaming effect.

8.8 Dryers

Oxidative drying, which occurs in addition to physical drying, is based on the reaction of oxygen with the double bonds of the resin molecules [13], of above all alkyd resins, but also with styrene-butadiene dispersions. Peroxide bridges, and finally carbon-oxygen or carbon-carbon bonds are formed from the initial oxygen-induced hydroperoxides. This reaction is visible by the gelation of the resin and finally the hardening of the film. The chemical reaction that takes place here has become technically useful because it can be greatly accelerated by the addition of metal soaps. The soaps concerned are octoates, decanates and naphthenates. The metal ions play the decisive role. According to metal ion, the following are differentiated:

- primary (above all cobalt and manganese ions which can exist in multiple oxidation states) and
- secondary dryers (calcium, zinc, barium, which only exist in one oxidative state) as well as so-called
- combination dryers (zirconium) that are active by a somewhat different mechanism.

What has been said thus far is valid above all for solvent-dissolved resins, especially since the resins are particularly soluble in non-polar media. It is valid in principle however, also for aqueous resins [91,113]. In the following, it will be shown that the circumstances of classical, solvent-containing binders can not be transferred completely to water-reducible systems [114,115]. Thus,for example, a transfer from the aqueous phase to the organic, hydrophobic binder appears on drying. At this point, it is impossible to give patent formulations for the use of dryers in water-borne coatings since the literature at present does not provide a cohesive picture. Possible problems that might occur will be discussed, however. To be aware of above all, are the interactions of the water-insoluble metal soaps with the components of a water-borne coatings formulation. The activity of a dryer can be reduced by hydrolysis, especially in the case of cobalt, or by complex formation either with water or with the amines present for neutralization (water solubility). The multi-valent cations (cobalt, manganese) can lead to binder coagulation, either with resin emulsions or also in hybrid systems of water-soluble resin and polymer dispersion, since emulsions and dispersions tend to be unstable with such ions. Even so pigment flocculation can be initiated by the presence of multi-valent cations. Combating these problems, which can arise with the use of the usual dryers dissolved in solvents, is being attempted by the use of emulsified dryers. [114]. These products are systems that contain chelate formers and emulsifiers and have a less negative effect on the resin. Good through-drying and gloss development is achieved with them even after prolonged storage.

It is reported that classical dryers should be incorporated before dissolving in water if possible [116]. With resin emulsions, incorporation before emulsifying is the most favorable. Admixture to the already completed emulsion is possible only with great difficulty. Emulsified dryers, on the other hand, can be mixed in quite normally.

In general, aqueous resin systems dry more slowly and display slower hardness development after extended storage. Gloss, too, is normally less if the resin has been stored. These effects must be minimized by selection of the correct dryer. In general, cobalt dryers do well here; in part, the presence of secondary dryers such as barium and calcium are of advantage. The selection of a classical dryer or an emulsified product must be made on a case by case basis. As a trend, the latter are advantageous, however.

In addition to the oxidatively-drying resins, the oxidative cross-linking of styrene-butadiene dispersions can also be accelerated [117]. This is to be recommended above all, when dispersions are used together with aqueous resins in hybrid systems. Combination dryers based on cobalt and secondary dryers have proven out here.

8.9 Corrosion Inhibitors

De-gassed salt-free water is known to not corrode iron substrates. If water-reducible coatings systems are applied, however, brown spots, attributable to corrosion products, frequently appear on the substrate. Obviously, certain components of water-borne coatings, such as salts and oxygen, promote corrosion. But, pH also has an effect, in that higher values usually causes less corrosion. This problem, generally known as "flash rust", is more than just an optical disadvantage – it can be the cause of subsequent corrosion. The effect can be prevented by the addition of corrosion inhibitors. These products must be distinguished from corrosion protecting-pigments, although both products can act very similarly [118,119). While the latter have the assignment of supporting the corrosion-protecting activity of coatings over an extended time frame, and thus can be only slightly soluble, the first are to be quickly, but only briefly, effective during application: They must have good water solubility.

An entire range of products has been known for some time that prevent flash rust. The products can be divided into four groups [120,121];

- oxidizing substances (chromates, dichromates, metaborates, sodium carbonate, sodium molybdate, sodium nitrite, sodium or potassium nitrate),
- organic substances (polar substances with active groups such as amines, carboxyl groups, sulfur groups, etc)
- metallic cations and
- non-oxidizing inorganic and organic salts (benzoates, naphthenates, octoates, etc.)

In addition to their effectiveness, environmental compatibility must also be considered. Thus heavy metal salts are today excluded (nitrates and nitrites are also currently under evaluation). With respect to the dosage, it must be noted that corrosion inhibitors, being salts and water-soluble compounds, can themselves injure the function of corrosion-protecting coatings by increasing water swelling. The results of salt spray and humidity exposure testing are to be considered in each case when determining the quantity to be used.

A theory of the mode of action of corrosion inhibitors has not been completely developed, since the relationships are relatively complicated [120, 122]. The key statement about the mechanism is that in the presence of salts of weak acids (e.g. sodium benzoate) the pH of the metal surface is set so that the iron is pacified by formation of oxidative layers. Similar effects are achieved by addition of nitrites, which also passivate the iron by reduction of dissolved oxygen to hydroxy ions. In this connection, it has been described that complex formers (tannins, acetylacetonates, salicylic acid, which for example are also used in rust converters) and certain emulsifiers can also have a favorable effect. On the one hand, they stabilize the oxidative pacifying layers of the iron while not unfavorably impacting the water swelling of the coatings. Emulsifiers to be sure, must not be added to the total formulation, but must be present in the binder. In this context, it must be mentioned the in the presence of corrosion inhibitors, the need for biocides is partially eliminated, since the former can have a similar retarding effect on the growth of microorganisms.

In conclusion, a problem must be discussed that is closely related to flash rust formation and is designated as "early rusting" [13, 123]. This phenomenon is of special concern when corrosion-protecting coatings, for example in relatively thin layers, are applied to cold substrates or at high humidity. The appearance is similar to that of flash rust, but is delayed. Apparently, this effect is attributable to the drying process, which proceeds from top to bottom, especially in the case of dispersion coatings. With slow drying, the inner areas of the coating remain decidedly wetter than the top due to the circumstances mentioned. Similar to the case of flash rust, corrosion, migration of iron ions and thus the formation of brown spots can result in the fresh state. Again, corrosion inhibitors are suited for prevention.

9 Pigments and Fillers

9.1 Introduction

Paints and coatings, independent of the nature of the binders, are inconceivable without pigments and fillers. While the pigments and their tinting function are in the spot light with typical water-based coatings, pigments and fillers are complementary in their effect in the customarily binder-poor classical dispersion paints [91]. In such products, fillers are not to be viewed primarily as "cheap", but much more they partially support the effectiveness of the pigment, influence the rheology, increase the protective effect of the coating and naturally also make the production of cost-effective serviceable paints possible [104]. At the latest, with the development of water-reducible corrosion-protection systems, it becomes clear how decisive the correct selection and composition of fillers is for the quality of the end product.

Fundamental differences between pigments and fillers for aqueous and solvent-containing systems do not exist. To be noted however, are the water-soluble components and the surface effects of the solid constituents of the formulation. Thus for example, soluble multi-valent cations can lead to stability problems with polymer dispersions, since the normally electrostatic stabilization system of the binder can react sensitively. The surfaces of inorganic fillers and inorganic pigments are usually hydrophilic, so that wetting with water need not first be effected with auxiliaries. A certain support is however necessary to stabilize a pigment in the dispersed state: For this, wetting and dispersing agents are used. In the case of organic pigments, the use of ready-made color pastes is sensible, since by this means, the paint producer is freed from the difficult dispersion of these pigments [91].

9.2 General Concepts

In this chapter, concepts that have general significance for the use of pigments and fillers will be explained.

An important parameter for describing the solid constituent content of a formulation is the pigment-volume-concentration PVC [81]. The name is actually not quite correct, since it applies not only pigments, but also to fillers. The PVC gives the volume-dependent proportion of the inert solids of the total volume of the hardened coating:

$$PVC = \frac{V_{Pig.} + V_{Fill.}}{V_{Pig.} + V_{Fill.} + V_{Binder}} \cdot 100\ [\%]$$

Naturally, the PVC can also be based on a special pigment or filler. This value thus describes quite clearly the relationships in a coatings material and permits direct inferences about some significant properties. Thus it is understandable that a water-borne coating with a PVC of 20% consists mostly of binder in which the pigment is imbedded. A coating made from this should have relatively low porosity and have a certain gloss. For an internal low-cost dispersion-based paint with a PVC of 85%, the binder content is so low that the solid components can just be held together. The total structure is correspondingly relatively open-pored and the appearance of the dried paint is matt.

An instructive parameter is the so-called CPVC, the critical pigment-volume concentration [124]. Up to this boundary concentration the binder, because of its volume proportion, is able to completely surround the inert constituents. It is understandable, that paints near this value display clear changes in their behavior which can be used for the experimental determination of the CPVC. Such test methods are for example the

- Gilsonite test (film porosity) [125],
- the measurement of film tension (maximum at CPVC) [81],
- the determination of the hiding power.

Of influence on the position of the CPVC of dispersions, in addition to the chemical composition, is the mean particle size: The smaller this is, the higher the CPVC and the higher the pigment binding capacity of the dispersion. The filler and pigment combination in a formulation also influence the position of the CPVC. The higher the binder demand of a filler, the lower the CPVC drops. Purely economic reasons usually favor a CPVC as high as possible, while technical arguments can favor the opposite. The binder demand of a filler mentioned above is described by the so-called oil number, especially in the context of classical resin systems. This indicates how many ml of linseed oil a filler will take up. For the water-reducible products of interest here, the water absorption or "wet point" is also a relevant value [104]. This describes, analogous to the oil number, how many ml of water a given quantity of filler can absorb. The two values do not run in parallel with diverse filler and pigments, which shows, that the effects differ. As already mentioned above, the wetting of fillers in aqueous media is generally not difficult, since the filler surfaces are generally hydrophilic, and thus water friendly.

The water or binder demand of a pigment or filler is dependent not only on the chemical composition or the surface morphology, but also on the packing density. This results, in turn, from the particle size, and since the particles are not uniform, from the particle size distribution. A given filler will take up less binder, the more densely it can be packed and the smaller the interstices between the particles are. It is evident, that effects on the applications properties such as hiding power, susceptibility to cracking, scrub resistance, rheology and film thickness, result from this and the particle size. For the analysis of particle size/particle size distribution, screen or sedimentation procedures as well as electron microscopy are used as a function of the median particle size.

Additional properties of pigments and fillers will be discussed in the following text. Criteria such as density, bulk density, hardness, water content and conductivity are only mentioned here, and will not be discussed further [91,126].

9.3 Titanium Dioxide

Today titanium dioxide is the white pigment. Other products play only a subordinate role. The overpowering significance of this pigment results from the great brightning effect and especially the highest opacity of all the white pigments [127]. With rutile and anatas, there exist two modifications of titanium dioxide, which differ in several properties.

Rutile is the more valuable product due to the higher opacity and the better weathering stability, while anatas has a purer white tint due to the stronger remission. Rutile pigments have a light yellow tinge. Remission spectra show that white pigments absorb practically no light in the viable range; much more, they scatter it [81]. The opacity effect results from this scattering effect: To be sure a white pigment returns all of the light, it prevents the penetration of light through the coating by this dispersion and thereby also the translucence of the substrate. The measure of the scatter can be determined for example, by the contrast ratio, in which the brightness of a white paint on a black substrate is compared with that on a white substrate. The more the index of refraction of pigment particles differs from that of the environment – for coatings systems primarily the polymeric binder – the more they can scatter light. If one notes that the indices of refraction for the usual resins lie around 1.5, it becomes clear why only substances with indices of refraction above 1.7 can be designated as white pigments. Herein is the significant difference between white pigments and "white" fillers: Fillers have indices of refraction of only 1.5 to 1.7, as shown in table 11.

Table 11 Refractive Indices of Fillers and Pigments in Comparison with Binders and Water [81]

Quartz	1.55	Zinc Oxide	2.00
Talcum	1.55	Zinc Sulfide	2.34
Kaolin	1.56	Anatas (TiO_2)	2.55
Calcium Carbonate	1.57	Rutile (TiO_2)	2.70
Mica	1.58	Alkyd Resin	1.53
Dolomite	1.6	Polyacrylate	1.48
Barium Sulfate	1.64	Water	1.33

Rutile and anatas, with indices of refraction of 2.7 and 2.55 respectively, lie substantially above this level and display the strongest opacity. In addition to the index of refraction, particle size, particle size distribution, pigment volume concentration and the degree of distribution of the pigment in the coating play a role. Regarding particle size, it is to be noted, that dimensions of 0.2 – 0.4 m yield the optimum coverage. This is connected with the wavelength of visible light, which is about twice as high. If the particles are smaller, opacity rapidly drops, since light can now quasi "go around" the particle. The dependence on pigment concentration makes definitely clear [128]: Only above a certain volume percent, can a white pigment manifest its effect. Thus in general, a PVC (based on TiO_2) of 10% is necessary to obtain an opacity of approximately 90%. With further increases in the pigment quantity, the opacity increases only slowly and can even drop after passing through a maximum. An optimal distribution in the binder is important for good utilization of a white pigment. Here on the one hand, it must be noted that the pigment particles must be satisfactorily dispersed, so that flocculation will not occur. The correct amount of wetting and dispersing agents should definitely be determined (see chapter 8.6). On the other hand, one can guarantee an optimal distribution by the addition of fillers that are as fine as possible and which act as spacers between the pigment particles [129]. If the pigment concentration is too high, the particles are no longer ideally distributed, and opacity sinks, as described above. To be noted is that the opacity of a paint changes during the course of the drying process [130]. Its development depends on various parameters such as titanium dioxide content, PVC and oil number of the fillers and also on the quantity of film builder. In general, the opacity of a paint below the CPVC will decrease on drying, while it will increase above the CPVC after passing through a minimum. The latter effect is also designated as "dry hiding". It is attributable to the fact that air is included in the film, which has a lower index of refraction than the water that was present before drying. Due to this replacement of water by air, the difference in the index of refraction of the individual components pigment, filler and binder to the environment (air) becomes greater than before drying.

Below the CPVC, by contrast, water is displaced by the polymer binder: The difference in the indices of refraction drops and likewise the opacity. There are however also possibilities to increase the opacity below the CPVC, by utilizing so-called plastic pigments. These are hollow spheres of polystyrene in dispersed form [131]. By this means, air is brought into the system, without lowering the film integrity. In addition, such polymeric fillers act as extenders for the pigment particles and thereby increase their effectiveness or enable a pigment savings. To be sure, they have no effect on the wet opacity of a paint which can only be increased by TiO_2.

Of great significance for the industrial applications properties of a TiO_2 pigment, as well as other pigments, is the surface treatment [132, 133, 134]. For this, SiO_2 and Al_2O_3 especially, but also ZrO_2, ZnO and other oxides such as small quantities of organic products (e.g. polyalcohols) are used. These additives influence the dispersability [135], the settling and flocculating behavior and the weathering resistance of the pigments and also effect viscosity, rheology and the optical properties of the paint produced. [136]. The weathering resistance of TiO_2 is very high, the pigment can however effect the photocatalytic degradation of the binder. On the one hand, TiO_2 protects the polymer in that it absorbs uv light, on the other hand, it can release electrons as a result of this energy take-up, which leads to film degradation (tied to loss of gloss, chalking) by formation of oxygen radicals or HO_2 radicals [137]. Interestingly, the negative effect of TiO_2 on uv-stable binders is greater than on susceptible systems. In the latter case the protective function of the TiO_2 dominates, since it keeps uv radiation away from the binder by absorption. In the first case, the degradation of the inherently stable system is induced by the TiO_2. This effect can however be countered by appropriate surface treatment of the TiO_2 [138, 136].

9.4 Color Pigments

It is not possible to present a comprehensive account of color pigments at this point [139]. The area is far to broad, so that only an overview can be given [140]. The most important individual products will be treated according to color tint. With respect to their chemical basis, color pigments can be divided into inorganic and organic products. Carbon black occupies a certain special position as a significant black pigment. The inorganic pigments in essence, are chemically inert, very light-fast products based on oxides and sulfides of the elements iron and chromium. Zinc, molybdenum and cadmium are of smaller significance. The last however, has become much less significant for toxicological reasons. Synthetically produced pigments are unambiguously dominant, since only they suffice for today's requirements for color consistency and uniformity. The various color tints result not only from the chemical composition, but can be varied by controlling the particle size and particle size distribution. The inorganic mixed-phase pigments are also an interesting group, in which the color effect is achieved by inclusion of foreign ions in the host lattice. Nickel titanium yellow, which results from the incorporation of nickel ions in the titanium lattice, is mentioned as an example. Among the most important representatives of the organic color pigments are those based on azo compounds, carbonyl colorants and phthalocyanines, as well as their salts or metal complexes.

There is no general rule, when which pigment group should be used. While inorganic pigments are especially distinguished by their high stability to chemicals and UV exposure, organic pigments, because of their fine particles, have greater tinting strength with lower opacity and greater luminosity. Further, the tint palette of the latter is significantly broader, so that organic pigments are frequently unavoidable, given the large significance of color tint fashions. In summary, it must be noted that the share of these products is also growing due to the ever-increasing quality. An important aspect for the user, next to the optical properties of a color pigment, is straight-forward incorporation in the paint formulation, maximum utilization of the generally expensive substances and a high color stability. These are achieved by a corresponding surface treatment and special micronizing techniques on the part of the pigment producer [132]. Optimal distribution of the pigment in the binder by use of serviceable dispersion aggregates, as well as a good determination of the wetting and dispersing agents to be used for stabilizing the pigments, is never the less essential. In recent years, pigment pastes have achieved increasing significance [91]. While they are nearly unavoidable with the use of organic pigments in aqueous systems, since an optimal dispersion with the customary aggregates of the color industry is very tedious, there is also an ever-increasing tendency to use such pastes with inorganic pigments. Without them for example, the modern computer driven color mixing systems, that function with utmost precision, would be inconceivable.

In the following, the most significant color pigments are presented arranged by color tint:

- Yellow pigments

  Iron oxide belongs to the standard pigments among the yellow pigments. Corresponding to the typical properties range of inorganic pigments, the luminosity and color tint range are relatively limited, with good resistance to weathering, uv light and chemical exposure. These pigments are used above all, in classical emulsion paints. While chromium and cadmium yellows are rarely used in aqueous media, the mixed-phase pigment nickel titanium yellow is especially suited. It has good opacity and due to its alkali resistance, like iron oxide yellow, it is also usable in silicate paints. The temperature stability is even higher than that of iron oxide pigments. Both types are available as aqueous color pastes. Of the organic yellow pigments, Hansa yellow (acrylamide yellow), which is distinguished by good weathering and chemical stability, is especially recommended [140]. Benzidine yellow and the valuable anthrapyrimidine yellow are also used.

- Orange pigments

  Among the inorganic pigments, chromium orange and also some cadmium orange types are especially noteworthy, while molybdenum orange is not suited. The iron oxides are also used in aqueous pigment preparations. As with the yellow pigments, acrylamide orange (Hansa orange) and diarylid orange (benzidine orange) dominate.

- Red pigments

  These are a clear domain of the synthetic iron oxide pigments, which cover the color tint ranges of orange, red and reddish brown. This pigment type is distinguished by high chemical and weathering resistance and well as good hiding power. Iron oxides are also used for brown tones, although the tones typical for umbra (also burned umbra) cannot be achieved. Pigment pastes are also almost exclusively iron oxide pigments.

Among the organic pigments, toluidine red (very high light fastness) and the very stable naphthol AS red are to be mentioned. In addition, perylene, thioindigo and quinacridone reds are especially used in water-reducible paints.

- Violet pigments

  For water-borne coatings, manganese and ultramarine violet from the inorganic range and China quinacradone and dioxazine violet from the organic palette are recommended.

- Blue pigments

  Ultramarine blue has the greatest light fastness of all blue pigments. It is used in emulsion paints and other aqueous systems, as is cobalt blue which is also a very light-fast chemically-resistant mixed-phase pigment (especially as cobalt aluminate). Among the organic products, phthalocyanine blue clearly dominates. This type is especially used for pigment pastes.

- Green pigments

  The chromium oxide pigments are unsurpassed with respect to acid and alkali resistance as well as light fastness. Besides these, some green mixed phase pigments are also recommended. Of the organic green pigments, only the phthalocyanines have achieved significance for water-based paints, especially in the form of color pastes.

- Black pigments

  Organic pigments play no role here [140]. Graphite is however mentioned as a possible pigment. Iron oxide products and diverse carbon black types however, have primary practical relevance. Especially in the case of carbon black, which is processed with difficulty, the use of pigment pastes is favored.

9.5 Fillers [13, 91, 126, 141]

With reference to the definition of white pigments in chapter 9.3., fillers are white or colorless inert solid products with an index of refraction of about 1.6. These substances contribute little or nothing to hiding power in the wet state or in paints with a PVC below the CPVC. Above the CPVC however, fillers also have an effect which has already been described (see chapter 9.3.) as the dry-hiding phenomenon. Colorlessness is a point of quality: A good filler should have as small an intrinsic color as possible, e.g. have a high degree of brightness, so as to not disturb the coloring effect of the pigment. This property results, since it is usually a matter of naturally-occurring products, essentially from the purity and fineness of the material, thus in the final analysis from the deposit conditions and the possible subsequent work-up steps. Synthetic fillers currently can contribute more, but to be sure, are correspondingly more costly. Optical behavior is however only one aspect of a filler. Of great significance, especially for water-reducible systems, is the chemical composition of a filler. In contrast to the coatings resins products dissolved in organic solvents, which are indifferent to inorganic fillers; in aqueous media under some circumstances, chemical interactions can occur. Thus the pH of a given filler in water is to be observed. Most products react alkaline, so that no problems arise. Since water-reducible paints are usually alkaline, that is when the best stability of the system is present. Some fillers, such as kaolin, precipitated barium sulfate and fumed silica, can however react acidic, which must be considered in formulating. As a result, a filler should be broadly stable to acids and alkalis and in addition, contain as few water-soluble components as possible, which could impair the stability of the binder. Especially in corrosion-protection systems, soluble components in fillers can lead to osmotic effects in the coating and thereby impair their integrity and protective function.

Table 12 Particle Structures of Fillers [104]

Form	Examples
Sphere, Sphere shaped	Glass Beads
Cube, Cubic, prismatic, rhombohedral	Calcit, Feldspat
Parallelepiqued, Tabular, prismatic, pinacoid, irregular	Calcite, Feldspar, Silika, Baryte, Nephelite
Leaflets, Platelet shaped, lamellar	Kaolin, Mica, Talcum, Graphite, Aluminum, Hydrate
Fiber, Fiber shaped	Wollastonite, Tremolite, Saw Dust

An important function of the filler is to increase the volume of the coating and thus improve the economics. This must not only be done without impairing quality, but should support product quality. The geometry of the filler, i.e. the size and the particle size distribution, plays a large role because of unavoidable variability. With respect to form, there are definite basic types to distinguish, which are shown in table 12.

Here too, one cannot assume a uniform structure, especially since the products partially forfeit their original form during grinding. Talcum and mica are mentioned as especially characteristic examples of lamellar fillers, which improve the film integrity and weathering resistance with appropriate orientation in the film and also improve the adhesion to the substrate. A further effect of lamellar fillers is their viscosity-increasing and thixotropic action, by which the rheology is influenced and the settling tendency of pigments can be reduced. In corrosion-protection formulations, the always critically observed thickener quantity can frequently be reduced by the clever use of lamellar fillers. In addition, lamellar products increase the dry-hiding effect and can effect matting. Fiber shaped fillers, such as wollastonite, and also wood or synthetic fibers increase the tensile strength of the coating be their reinforcing action. At the same time, they can act as thixotropes or thickeners.

For the above-mentioned main function of a filler, the increase in volume, the mean particle size and the particle size distribution play a definitive role. These properties are determined primarily by the producer by the preparation process (grinding, screening). In this regard, it is of crucial significance to the user that these values be subject to the narrowest supply specification ranges, especially when one realizes, that in the usual paint formulations, combinations of different fillers are used. The particle size distribution especially, is closely related to the packing density of a filler. A further criterium is the specific or inner surface, which results from the geometric form and especially from the surface quality. The larger the specific surface, the stronger the interactions with the usual formulation components, especially with the binder and the solvent water. Finally, the binder demand of a filler, which is customarily characterized by the previously described oil number or wet point, results from the packing density and specific surface. For cost reasons, the binder demand of a formulation should be as small as possible.

An additional aspect, which is also intimately tied to the surface structure of a filler, is the wetting behavior. Since ground products are hydrophilic due to the OH groups on the surface, wetting by water, in contrast to solvent-containing systems, is not difficult. Never the less, most formulations do not obviate the need for auxiliaries. This has the following cause: After wetting, the filler agglomerates, that are always present, must be disintegrated, and after attaining an optimal disintegration and distribution, this state must be stabilized. By stability is meant the prevention or a reagglomeration or flocculation of the dispersed fillers and pigments by wetting and dispersing agents. These enclose the individual particles and thus prevent a renewed accumulation, which would reduce the effectiveness of the products. The optimal additions of wetting and dispersing agents must be determined experimentally, since they are specific for each individual pigment-filler-binder combination. Both excessive and insufficient doses bring disadvantages.

Additional properties of fillers are taken up in the following individual discussions.

9.5.1 Silicates

Silicious products represent the largest single group within the filler family. In essence only the chemical basis is common to all while the precise structure, specific properties and thus the application areas differ greatly. To be mentioned as key words here, are natural and synthetic silicic acids and silicates in the true sense, e.g. salts of silicic acid, such as talcum, mica and kaolin.

The quartz powders, which originate from natural deposits and which are dressed only by purification, grinding and screening, as well as the narrowly-used cristobalite, which is largely produced synthetically, also belong to the silicic acids. These products are distinguished by favorable pricing as well as high mechanical and chemical resistance.

Cristobalite differs from the generally color-neutral quartz powders by its even higher degree of whiteness and is used with classical architectural paints and plasters, especially in road marking paints, where its hardness increases the wear resistance. In addition to the standard products, there exists a range of surface-treated types, which are in part suited for water-reducible coatings substances.

In the group of synthetic silicic acids, the fumed products are no longer considered as true fillers, since they are used as thickeners or rheological additives (thixotropic agents). They were already treated in chapter 8. Precipitated silicic acids and the related silica gels are essentially used for matting and to control rheology as well as in combination with TiO2 to increase the degree of whiteness and coverage, and thus optimize the pigment usage. As has already been mentioned, the incorporation of air bubbles (dry hiding) and the improved pigment distribution (spacer), play a role in the latter effect. Blocking resistance and susceptibility to scratching are also improved by the addition of silicic acids. In general, the price level of these products is in part substantially above that of "mere" fillers.

In addition to the synthetic products, natural silicic acids such as kieselguhr are used, which are usually dressed by calcining [91]. They are also used for matting, whereby for example, the disturbing sheen of interior paints, which occurs occasionally, can be eliminated.

Talcum, as magnesium silicate, is a typical representative of the silicate family. The lamellar structure is of quite decisive significance for its application. Because of this, and the higher oil number compared with spherical filers, talcum effects the viscosity and rheology of coatings. A structural viscosity behavior usually occurs. In comparison with mica, the reinforcing effect is not as pronounced, however talcum reduces the susceptibility to cracking. In typical interior paints, the scrub resistance is increased and in corrosion-protection paints, the protective effective is improved by lengthening the diffusion path. Similar to what was described for precipitated silicic acids, talcum acts as an extender for TiO_2 pigments and thus increases their effectiveness.

Mica is closely related to talcum, based on its chemical composition. Silicon is partially replaced by aluminum in its crystal lattice. The similarity is also expressed in the lamellar structure. As a whole, mica is rated more valuable than talcum. Its chemical and weathering resistance is very good, and it improves the light stability of the total system by means of a certain absorption in the uv range. Its reinforcing action is more pronounced compared with the relatively soft talcum, provided the lamellar structure has not been partially destroyed by grinding (depending on the dispersion process). Mica is surpassed in this respect only by the synthetically obtained special product plastorite [91], which is obtained by coalescence of mica with quartz and chlorite. For that reason, similar effects are observed on adding mica to a formulations with talcum: increased corrosion protection, increased surface hardness and thereby better scouring resistance and more effective pigment use.

Kaolins, which are frequently also referred to as China clays, are aluminum silicate hydrates. They also display a lamellar form. Especially pronounced is the construction of a "house of cards" structure in aqueous media due to the differing charge distribution: The edges of the platelets have a positive charge, the surfaces a negative one. With the use of large quantities, this effect can lead to a gel that is no longer capable of flowing. This can however be suppressed with wetting and dispersing agents. In general one can thus induce thixotropy and improve the anti-settling behavior by the addition of kaolins. Calcined clays show no rheological effects, but can be used like untreated clays for matting and to increase hiding power and the degree of whiteness analogous to talcum. The weathering resistance of kaolin is not as high as that of mica. Use in corrosion-protection-formulations is possible only after neutralization with amines due the acidic pH. Kaolins play a significant role especially in aqueous paper colors.

Precipitated aluminum silicates are characterized by a high degree of whiteness and, because of the pronounced fineness of the particles, a high oil number. They improve pigment dispersion, have a matting effect and lower the critical pigment volume concentration, so that they increase the porosity of a coatings system.

Wollastonite is a fiber-formed calcium silicate, which is also in part produced synthetically. To some extent, it can be used as a replacement for asbestos. The reinforcing action for reducing cracking susceptibility, as well as a certain thickening and thixotropic effect, is pronounced. At this point it must be mentioned that there are further specially-modified, silicious fillers, which because of their special structures, are suited for special applications such as sound and heat insolation and for extremely light coatings.

9.5.2 Calcium Carbonate

Next to the silicates, the calcium carbonates certainly represent the most important filler group. As with the silicates, one can differentiate between synthetic and natural products, whereby the latter have by far the largest share. The abbreviations CCN (calcium carbonicum naturale) for the natural and CCP (calcium carbonicum praecipitum) for the synthetic are frequently found. Of direct biological origin are the chalks, whose name is frequently used incorrectly as a synonym for the entire calcium carbonate group. The term chalk, in a rigorous sense however, refers only to sedimentary rock of biological origin in which the remains of tiny crustaceans can be observed under the microscope. In contrast calcite, the other CCN representative, consists of crystalline calcium carbonate and shows absolutely no crustacean remains. Marble and limestone powders are the most significant. A further difference is in the $CaCO_3$ content, which should be at least 85% for chalks and 99% for calcites. Synthetic calcium carbonate CCP is produced by treatment of slaked lime $Ca(0H)_2$ with purified carbon dioxide CO_2.

The great significance of the calcium carbonates results from their "unlimited" utility, their favorable price and good technical properties. Thus they have become the most important filler period in the paint realm. Of the positive properties, the physiological harmlessness, the high degree of whiteness, the low binder demand, the good dispersability, and as a result of the relatively low hardness, low machinery wear, are emphasized. With respect to stability, notice must be taken of the poor acid resistance. The passivating action resulting from the alkaline reaction is utilized above all in corrosion-protection paints: Calcium carbonates thus support a chemically induced anticorrosive effect. The synthetic calcium carbonates, by virtue of their significantly higher price alone, do not represent an alternative to the natural products, but much more, a combination of both products is opportune. By virtue of the particle fineness and greater brightness, precipitated calcium carbonates can contribute to the optimization of the pigment-filler proportion and thus the price. By their use, the hiding power and pigment distribution can be improved and the disadvantageous sheen in highly-pigmented paints can be suppressed by matting. In summary, a formulation- specific solution must always be determined for using CCN/CCP combinations.

Closely related to calcite is dolomite, a calcium-magnesium carbonate, which has a somewhat higher resistance to acids.

9.5.3 Additional Inorganic Fillers

In addition to the silicates and carbonates, other inorganic fillers of different chemical bases are used. They do not possess the high significance of the first named, but they will be briefly mentioned.

Aluminum hydroxides are used above all as physiologically harmless flame-proofing agents in dispersion systems (paints, plasters, fillers, adhesives). Micaceous iron consists in essence of iron oxide and has a platelet-shaped structure. It has high resistance to uv light and weathering. The primary application areas are corrosion-protection paints. Barium sulfate is used as a natural filler (barite) and as a precipitated product. Its significant advantage, next to good weathering properties, low oil number, easy dispersability and good gloss development lies above all, in its high specific density, which is of significance in weight-based packaging.

9.5.4 Polymeric Fillers

The adjective "polymer" serves above all to differentiate against the inorganic products and to give an indication of the raw materials basis. Intended here, are on the one hand, fiber-shaped products of cellulose or synthetic base or on the other hand, the so-called plastic pigments consisting of polystyrene [131]. The latter were already mentioned in chapter 9.3. They serve essentially to save titanium dioxide, whereby this affects only the dry coverage and not the wet coverage. Fiber-shaped fillers exist in the most varied manifestations. Especially cost effective are cellulose products, which are produced from wood or cotton. Valuable with respect to stability and property range are the synthetic fibers based on polyamide, polyacrylonitrile, polyethylene and polyoxymethylene. These synthetic products are offered as smooth cut grades as well as highly fibrillated grades [140]. In general they serve mechanical strengthening and thus are effective against cracking. As fibers, they can also influence rheological properties.

9.6 Corrosion-Protection Pigments

Since corrosion inhibitors were already treated in chapter 8.9., the corrosion-protection pigments will be discussed here. Both concepts describe the different applications goals only imprecisely. Both product groups in principle have the same task, namely to prevent the formation of rust. They differ however in the time frame of their activity. As inhibitors are understood products that prevent the formation of rust only temporarily, and specifically during the application of a water-borne coating and the subsequent drying phase. The type of rust occurring here is also designated as flash rust. Corrosion-preventing pigments, on the other hand, act permanently and are used not only in aqueous corrosion-protection systems, but are also used in solvent-containing products. The mode of action is quite similar. Because of the differing demands, there is a significant difference between corrosion-protection pigments and inhibitors especially in their water solubility. Inhibitors must be quite soluble to be able to be effective immediately. For the corrosion-protection, pigments a certain solubility should be present on the one hand, so that they in fact can display a chemical activity. On the other hand, this solubility should be only slight, since otherwise it could lead to osmotic effects and thereby to blister formation and deterioration of the corrosion protection.

If the relevant literature is reviewed [141], while chromates are frequently mentioned as ideal, they are no longer promotable for environmental reasons, and alternatives to them are indicated. Chromates show a very good effect as a result of supplementary chemical and electrochemical processes. Due to their quite high solubility, they give good results especially in the early phases of a coating. This behavior is especially advantageous for dispersion systems, which as is well-known, have weaknesses in the fresh state. In the case of aqueous epoxy systems, however, shortcomings resulting from the high solubility have been reported [143]. In summary, one must assume, that chromates today must be, and can be, replaced by other products, if aqueous systems are to achieve reduced environmental pollution in all respects.

The products to be discussed in the following are roughly differentiated as electrochemically, chemically and purely physically active substances, although optimal protection often is achieved through a combination of differently active pigments.

Phosphates, especially zinc phosphate, play a large role as replacements for chromates. The mechanisms of their activity are still not fully elucidated [144]. Certain however, is that phosphates prevent the solution of the iron in the anodic region by a passivation with formation of complexes of phosphates and oxides. Molybdates and tungstonates display a similar effect. In the case of zinc phosphate, zinc acts as an inhibitor in the cathodic region by the formation of sparingly soluble zinc hydroxides or basic zinc salts [145].

$$(H_2O + 1/2\ O_2 + 2e^- = 2\ OH^-)$$

As already mentioned, it is of decisive significance to the activity of a corrosion-protection pigment that has a certain solubility. For zinc phosphate, this occurs only in the acidic region. Correspondingly, it is reported that coatings systems with zinc phosphate function better in an acidic industrial atmosphere than in rural areas. As a result, Zn phosphate has a delayed effect, since the hydrolysis of zinc hydroxide and zinc hydrogen phosphate proceeds relatively slowly [146]. This makes clear that mixtures with other active substances is necessary. Borates and metaborates, which find application by themselves as corrosion-protection pigments, are to be especially mentioned here. Their advantage lies above all, in their higher solubility, which enables activity in the early stage. Pigment properties (particle fineness, form) are further optimized by modification with zinc oxide, molybdates, aluminum and also organic products [147, 148]. Developments should not be considered complete here [149]. The goal of all variations is attaining high activity from the early stages to the "old age" of a coating, thus a continuously present, yet not to high solubility. Not to be forgotten here is that the binder must support these processes by a certain swelling [147]. A fully impervious system could not provide this.

The latter is understandably also valid for the more chemically active corrosion-protection pigments. These products display activity in that they react with the binder, the metal surface and/or corrosion promoting substances that have penetrated, and thus make them harmless. To be mentioned as most important, are borates and metaborates, which establish a high pH in the coating by capture of acidic products (ions, film degradation products) and are also capable of forming passivating protective layers in the anodic region [150]. The relatively high solubility of borate can be reduced by SiO_2 modification. Zinc oxide reacts in similar fashion. Silicates, that are capable of an ion exchange reaction in the coating, are an interesting group [151]. Cations that have penetrated are immobilized and freed calcium ions diffuse to the metal/coatings substance boundary where they build up an inhibiting protective layer. Simultaneously, the penetration of corrosion-promoting anions is prevented. Advantageous here is that heavy metals can be completely avoided; disadvantageous is the relatively long induction phase. Products of this type offer good extended activity [152]. Organic and inorganic salts, to be sure, are also mentioned as corrosion-protection pigments; in water-reducible systems they act more as inhibitors.

In summary it must be established, that a corrosion-protection pigment or a combination of such products must fulfill the following requirements [120]:

- Active in a pH range of between 4 and 10,
- Slight, but sufficient solubility,
- Inhibition of both cathodic and anodic corrosion reactions
- Formation of poorly soluble products on the metal surface
- formation of an inhibiting layer between metal and coating without detriment to adhesion.

A further group of corrosion-protection products is tied to the key word adhesion. Zirconium aluminate and titanate improve adhesion, especially wet adhesion, to metal, by formation of covalent compounds [153,154]. Since as is well known, corrosion is always tied to a loss of adhesion, such products can have a positive affect on the coatings thus modified.

The substances mentioned so far, act on the basis of chemical processes. To do so, the products utilized must develop a certain activity in the form of soluble substances. Corrosion-protection products, that act in purely physical fashion by the barrier effect, follow a completely different concept. These are thus also designated as passive pigments. The barrier effect consists in the significant lengthening of the diffusion path for corrosion-promoting agents. Platelet-shaped fillers such as micaceous iron, mica and talcum are prominent. In the meantime, synthetic platelet-shaped iron oxide pigments have also been offered which display a higher blocking action than the customary pigments [155,156].

In general, it must be said that this process can function only in combination with an exceptionally impervious, non-swelling binder; that in addition, a very good adhesion to the substrate must exist. Only so can the diffusion of ions and gases be effectively blocked. To be noted here, is the fact that no absolutely impervious, e.g. in practice unattainable, ideal binder layer is needed. It is sufficient that the aggressive substances cannot exceed a certain threshold concentration, by reduction of the diffusion process. This results in the fact that corrosion processes, which like all chemical processes are concentration dependent, do not proceed.

Current practice might be based, in general, on a combination of the three methods that have been described (corrosion protection by active pigments, improvement of adhesion and the barrier effect). The latter process is certainly attractive due to its harmlessness with respect to the growing demands for environmental protection.

10 The Action of Amines and Auxiliary Solvents in Aqueous Baking Enamels

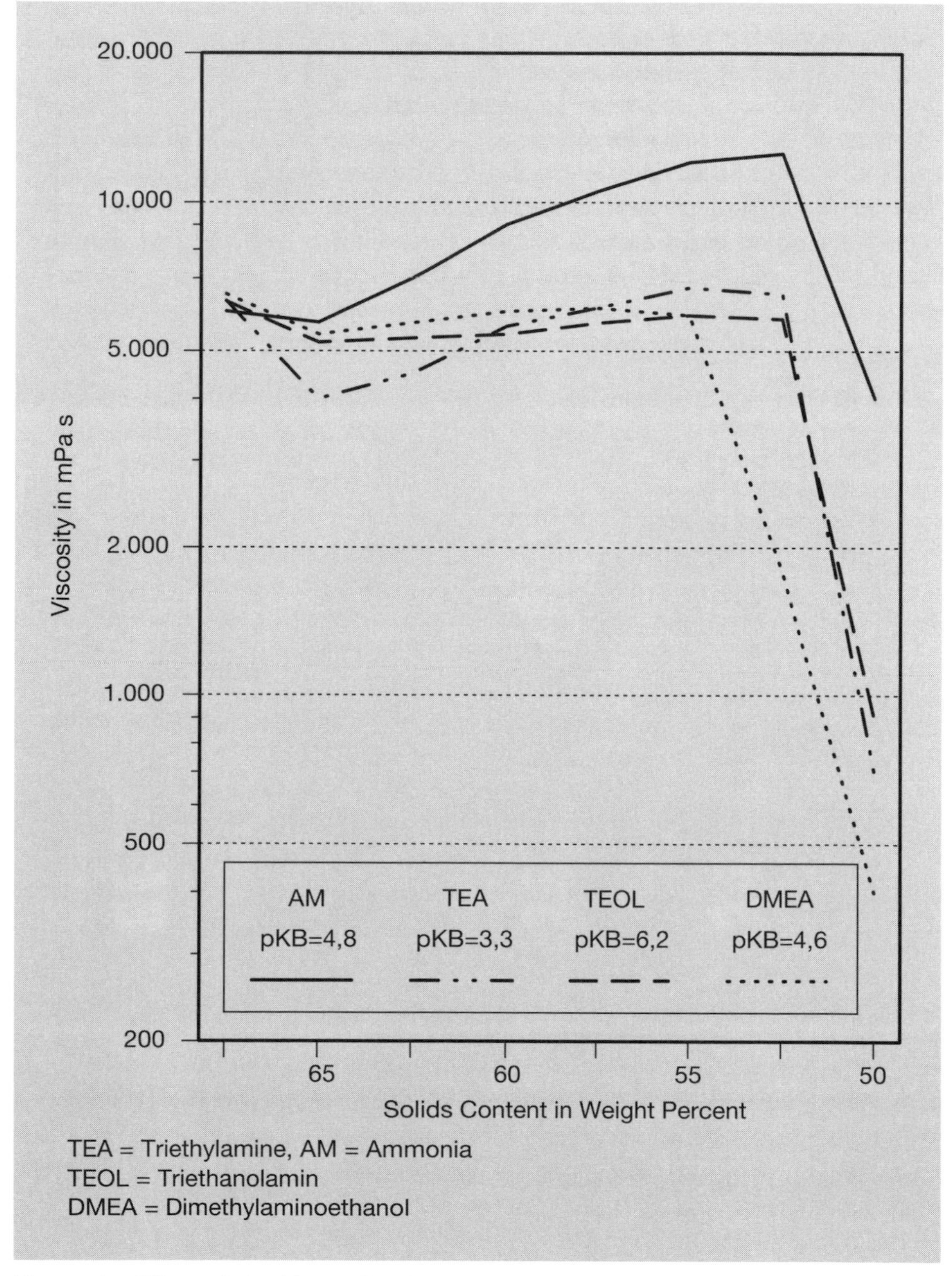

Figure 5 Effect of the Neutralization Agent on the Dilution Curve

The following investigations were carried out for the example of saturated polyesters bearing carboxyl and hydroxyl groups, which after neutralization with amines, form binder salts and are customarily used as baking enamels with melamine-formaldehyde resins [42].

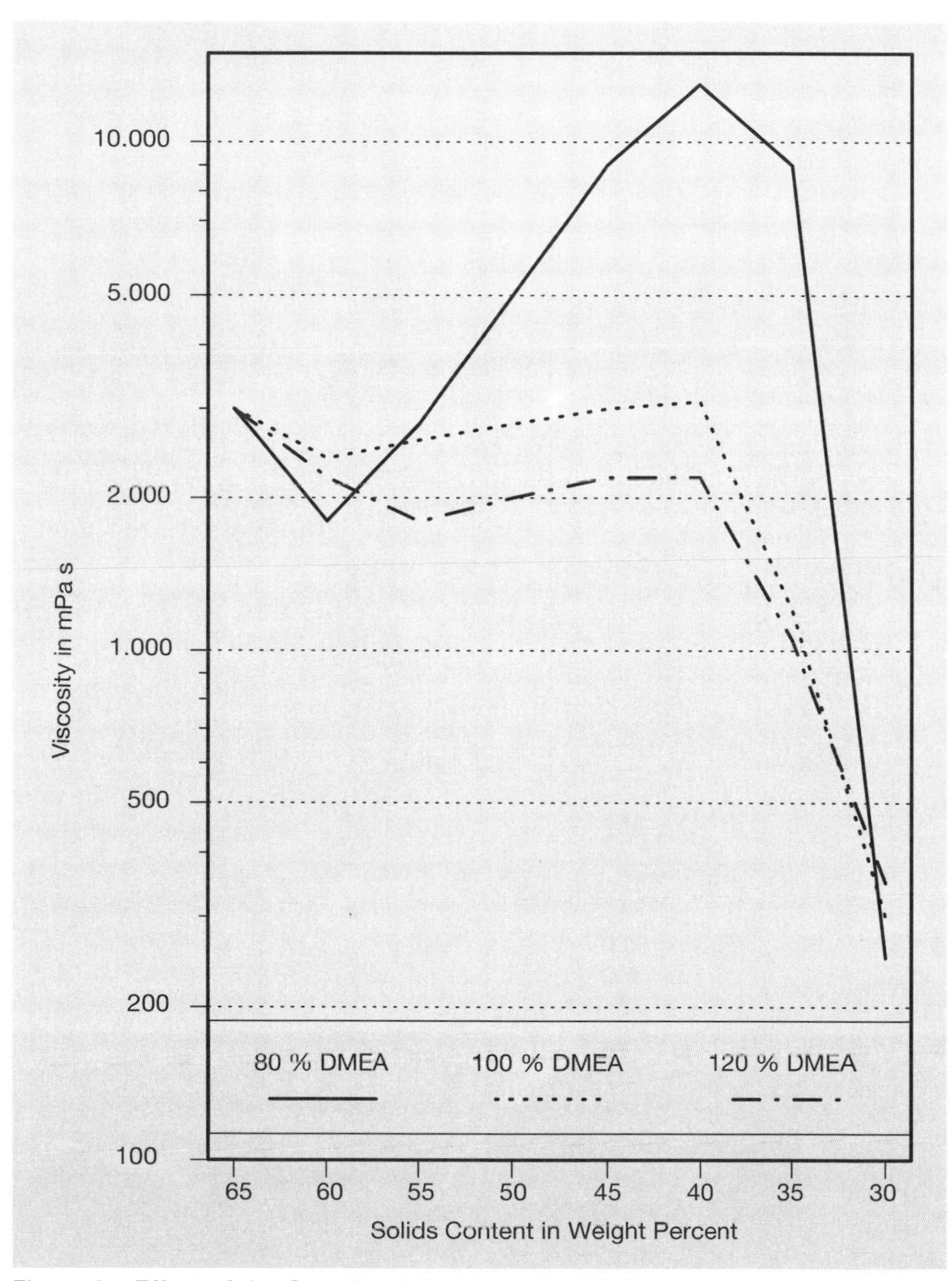

Figure 6 Effect of the Quantity of Amine on the Dilution

10.1 Effects of the Neutralizing Agent

For the neutralization of acid group-bearing binders with amines, those products should be selected that, in addition to the neutralization activity, have a certain solvent character (see also chapter 8.1).

To these belong the aminoalcohols, compounds such as dimethylaminoethanol or dimethylaminopropanol, which convey better reducibility with water and better storage stability [66, 67]. The effect of various amines on the dilution behavior is shown in figure 5.

All substituted amines are more effective than ammonia. There is no discernable relationship between the effectiveness of the amine on the viscosity behavior of the solution on diluting with water and the base strength – expressed as the basic constant. This is shown by a comparison of ammonia with dimethylaminoethanol, which have almost identical pKB values, but which differ by a large "viscosity peak" on dilution. Triethylamine, as the strongest base, has a less favorable effect on the coatings viscosity than triethanolamine, the weakest base in the group investigated. This observation permits the conclusion that amines, in addition to forming salts, act as auxiliary solvents. It is to be noted that triethanolamine, despite its good viscosity-suppressing properties, is used only under exceptional circumstances in the coatings sector due to its low volatility.

Figure 6 shows the effect of the quantity of added amine on the viscosity behavior upon dilution with water. To suppress an increase in viscosity, a 100% neutralization, based on the acid number of the binder, should occur; the effect of over-neutralizing the binder is unimportant from the stand point of viscosity. With 80% neutralization, the salt is less hydrophilic and thus less soluble in the concentration range in which intermolecular binder associations occur.

10.2 Viscosity Anomaly

As a rule, acidic coatings resins can not be thinned with water alone after neutralization with amines. They need the addition of auxiliary solvents, which function as solubilizers (see also chapter 8.2.). However a constant drop-off in viscosity, as is known for organic binder solutions, does not occur on dilution with water.

More frequently, upon diluting aqueous binder solutions, the viscosity increases sharply after an initial drop-off, and then after crossing through a maximum, again drops steeply. The significance of this viscosity anomaly, which is undesired for coatings practice – already mentioned in the portrayal of the amine activity – arises as follows:

Upon the addition of water (figure 7) to the solution of the polycarboxylic acid amine salts, two competing effects occur. On the one hand, dilution occurs which results in a reduction in viscosity. On the other hand, the solvating power of the mixture of auxiliary solvent and water is reduced by the addition of water. An association of the binder molecules results, which effects an increase in viscosity. The viscosity passes through a maximum in which the association is concluded [157, 158]. Upon further addition of water, only thinning occurs, which results in a sharp drop in viscosity, as is known for emulsions in the high solids range. Amines and auxiliary solvents must be selected so that the height and position of the viscosity maximum in the concentration diagram does not disturb the industrial coatings properties. Correct selection of the amines drops the viscosity significantly.

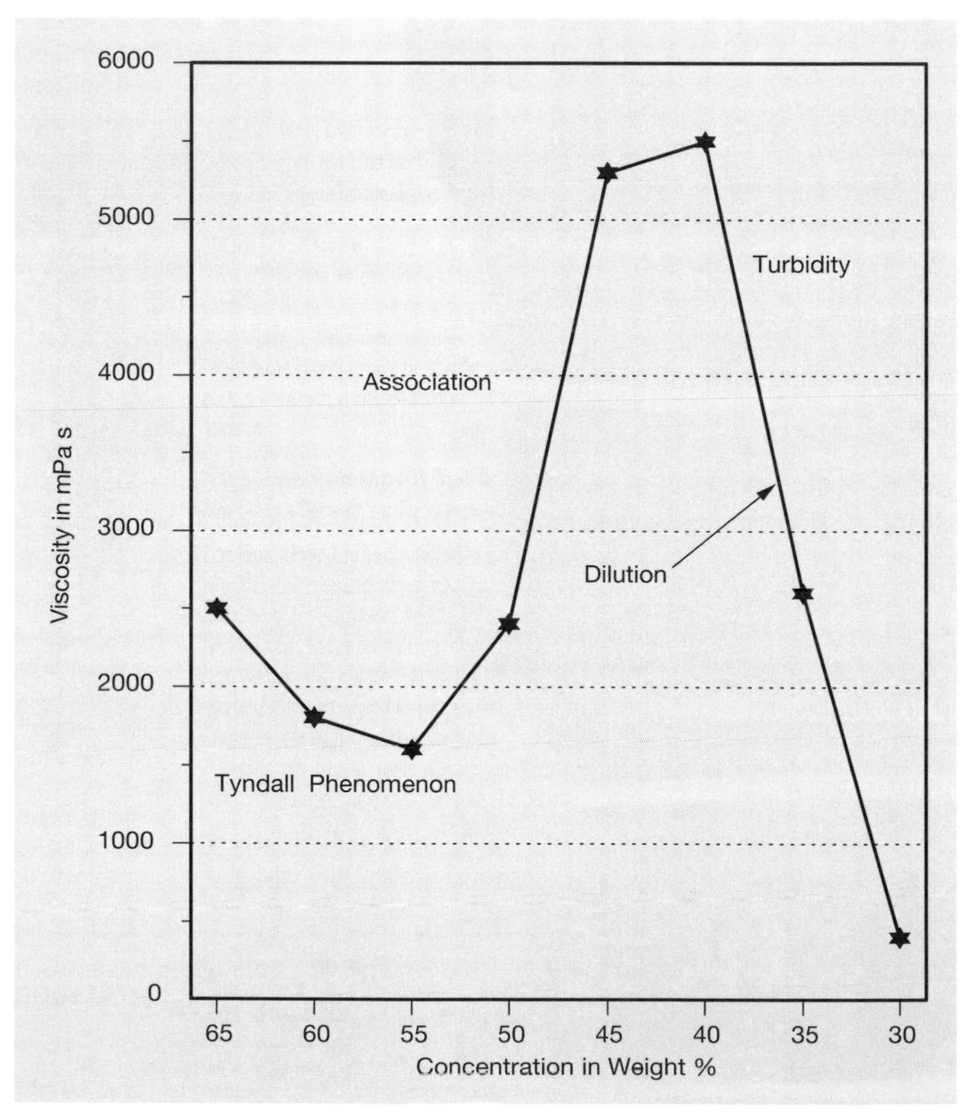

Figure 7 Viscosity Anomaly upon Thinning of Water-Borne Coatings

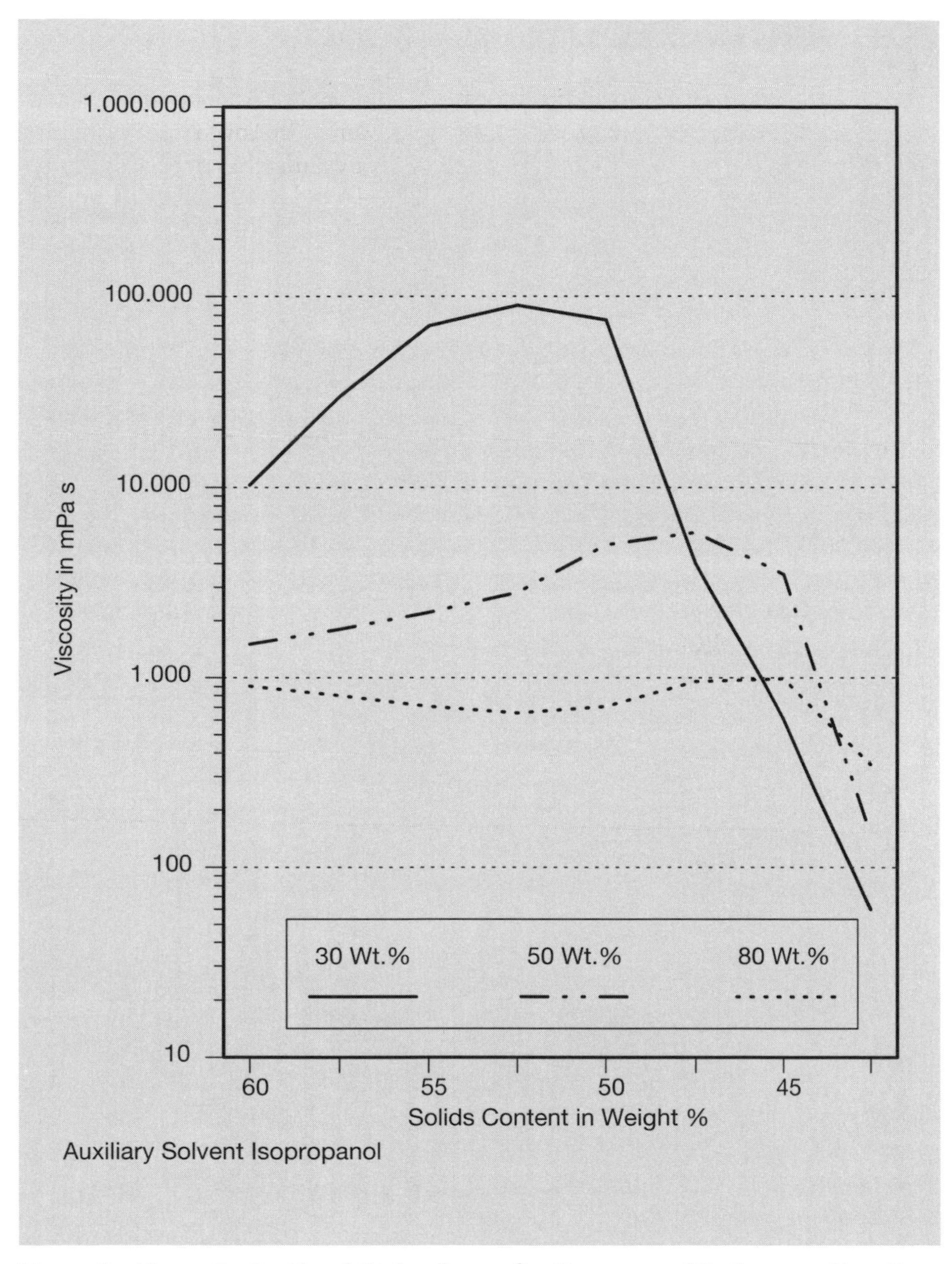

Figure 8 Viscosity Profile of Water-Borne Coatings upon Dilution as a Function of the Quantity of Auxiliary Solvent

10.3 Effects of the Auxiliary Solvent

By increasing the content of auxiliary solvents, the viscosity behavior of an acid polyester neutralized with amines can be made to approach that of a solvent-containing conventional system and thereby the industrial coatings pro-

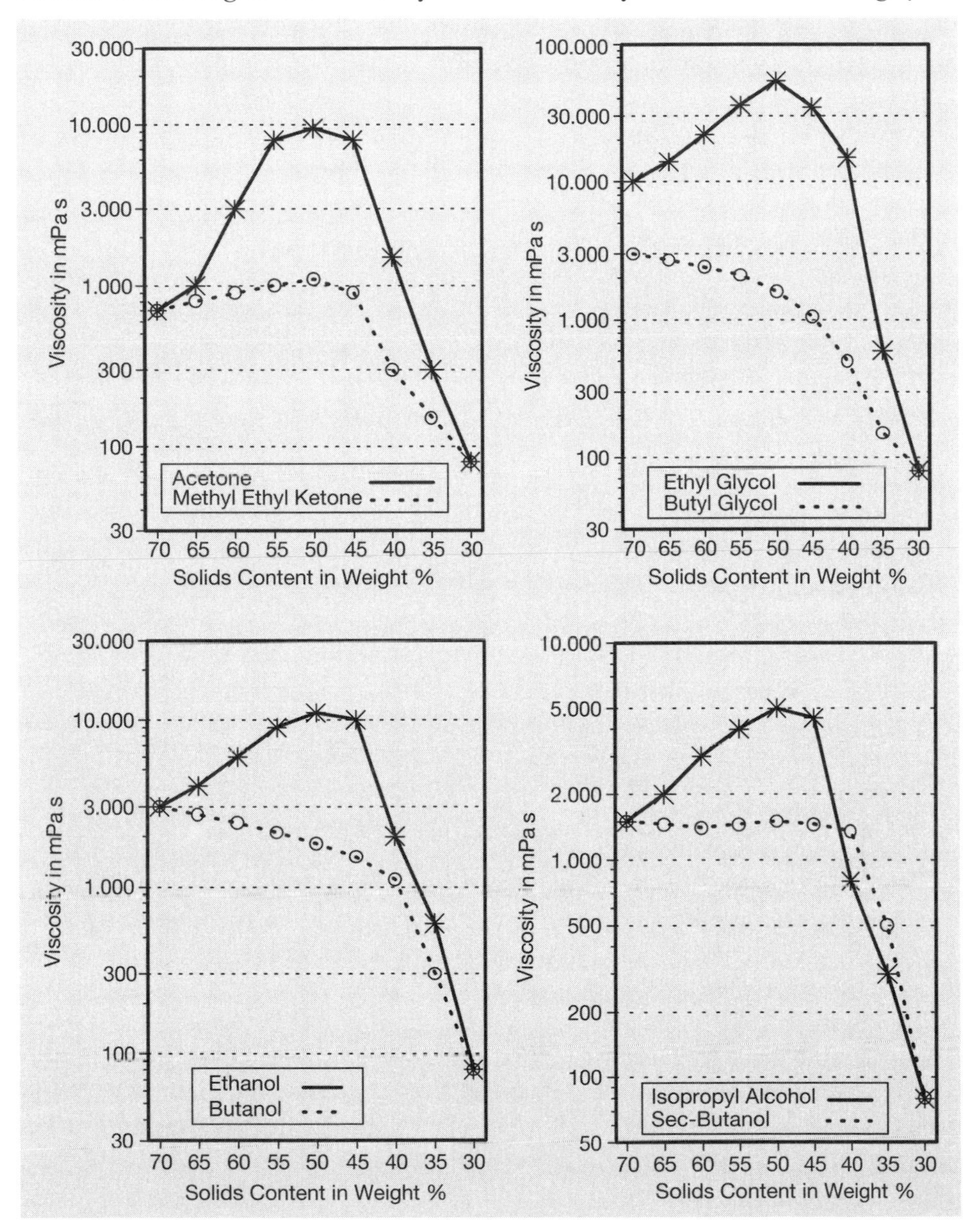

Figure 9 Dilution Curves of Water-Borne Coatings as a Function of the Type of Auxiliary Solvent

perties can be significantly improved (figure 8) (see chapter 8.2). The reason for this is that as a result of the increased solvent content, the solvating power of the resulting mixture of water and organic solvents is increased and thus the viscosity-increasing association of the polyester is suppressed. To be sure, it is not suitable to force a favorable viscosity behavior of aqueous coatings solutions by increasing the solvent content, since the increased organic substances content results in increased burdening of the environment on processing the coating.

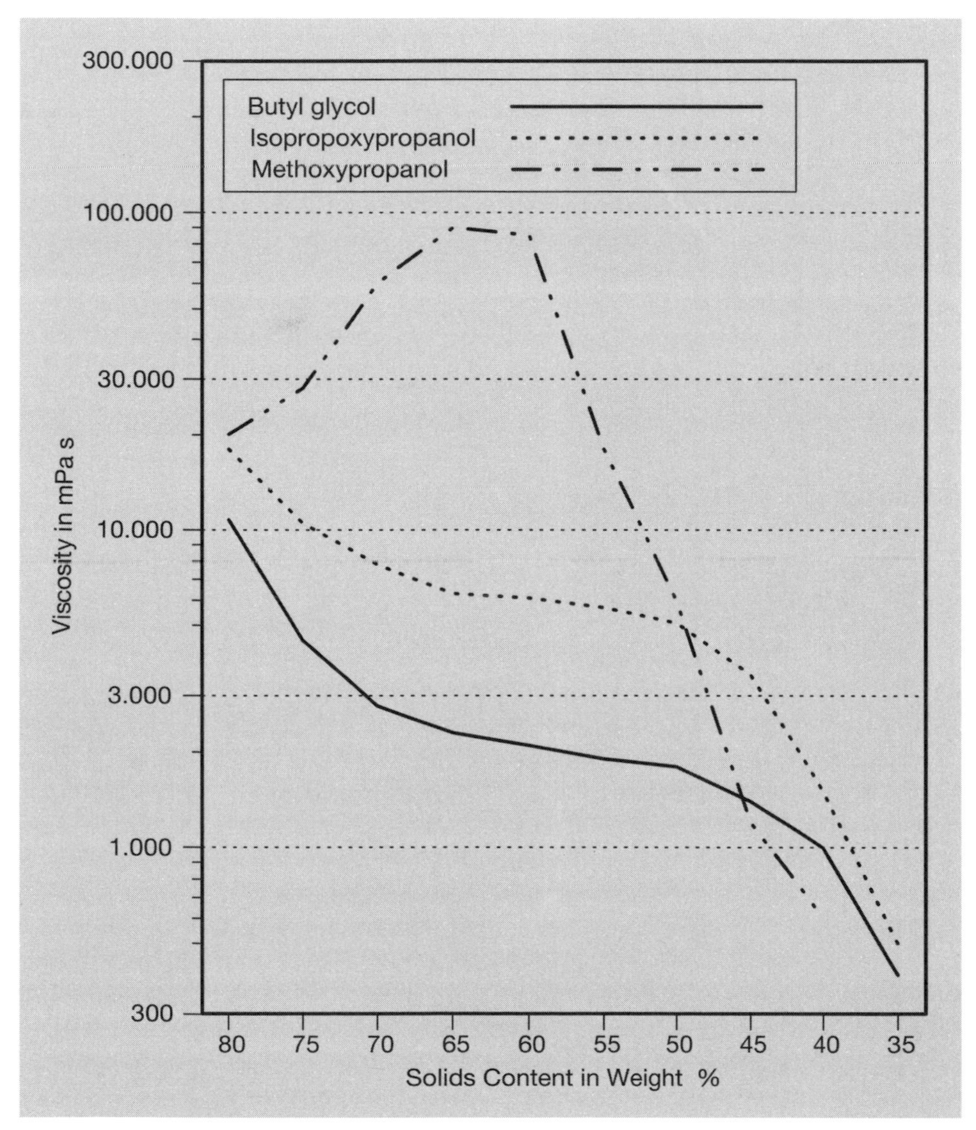

Figure 10 Influence of Diverse Glycol Ethers on the Dilution Behavior of Water-Borne Coatings

The problem thus arises of how to make the viscosity behavior of an aqueous coatings solution approach the familiar dilution behavior using small amounts of auxiliary solvents – how therefore, to lower the "viscosity peak" with only small additions of auxiliary solvents.

Ladder studies have shown that not only the amount of auxiliary solvent, but also its chemical structure, determine the height of the viscosity maximum [42]. Solvents with longer alkyl groups (see figure 9), as for example butanol or butyl glycol, result in more favorable dilution behavior than their lower homologues with methyl or ethyl groups, as for example methanol, ethanol or ethyl glycol. Similar, if not quite as pronounced, effects are found in the comparison of acetone with methyl ethyl ketone. The following explanation is offered for this phenomenon: The longer alkyl groups in the auxiliary solvent are more hydrophobic than the shorter ones and therefore have a greater affinity for the dissolved resin. Auxiliary solvents with longer alkyl groups orient themselves preferably with the resin molecules. The result is lowered association of the resin molecules with each other and thereby a flattening of the viscosity peak. Higher solvent homologues possess a quasi-surfactant character [69]. The lower solvent homologues, on the other hand, orient themselves to a large extent, with the aqueous phase due to their increased hydrophilic character, and therefore are not in a position to favorably influence the polymer association. Thus is explained why butanol, although of only limited miscibility with water, acts as a good auxiliary solvent in water. As a result of its orientation with the polymer, butanol has completely altered its familiar solvent properties.

The effectiveness, as auxiliary solvents in water-borne coatings, of the mono ethers of propylene glycol, which are offered as alternatives to the ethers of ethylene glycol, has also been investigated (see figure 10). Methoxypropanol and ethoxypropanol are not able to reasonably reduce the viscosity peak in coatings. Isopropoxypropanol and n-propoxypropanol behave significantly better in this respect. The viscosities of the coatings in general, lie on a higher level, so that in many cases, larger quantities of solvents are needed than with the use of butyl glycol. Butyl glycol, of all the solvents, obviously has the best properties in water-borne coatings, that until now have been attained by no other solvent [68,70,72].

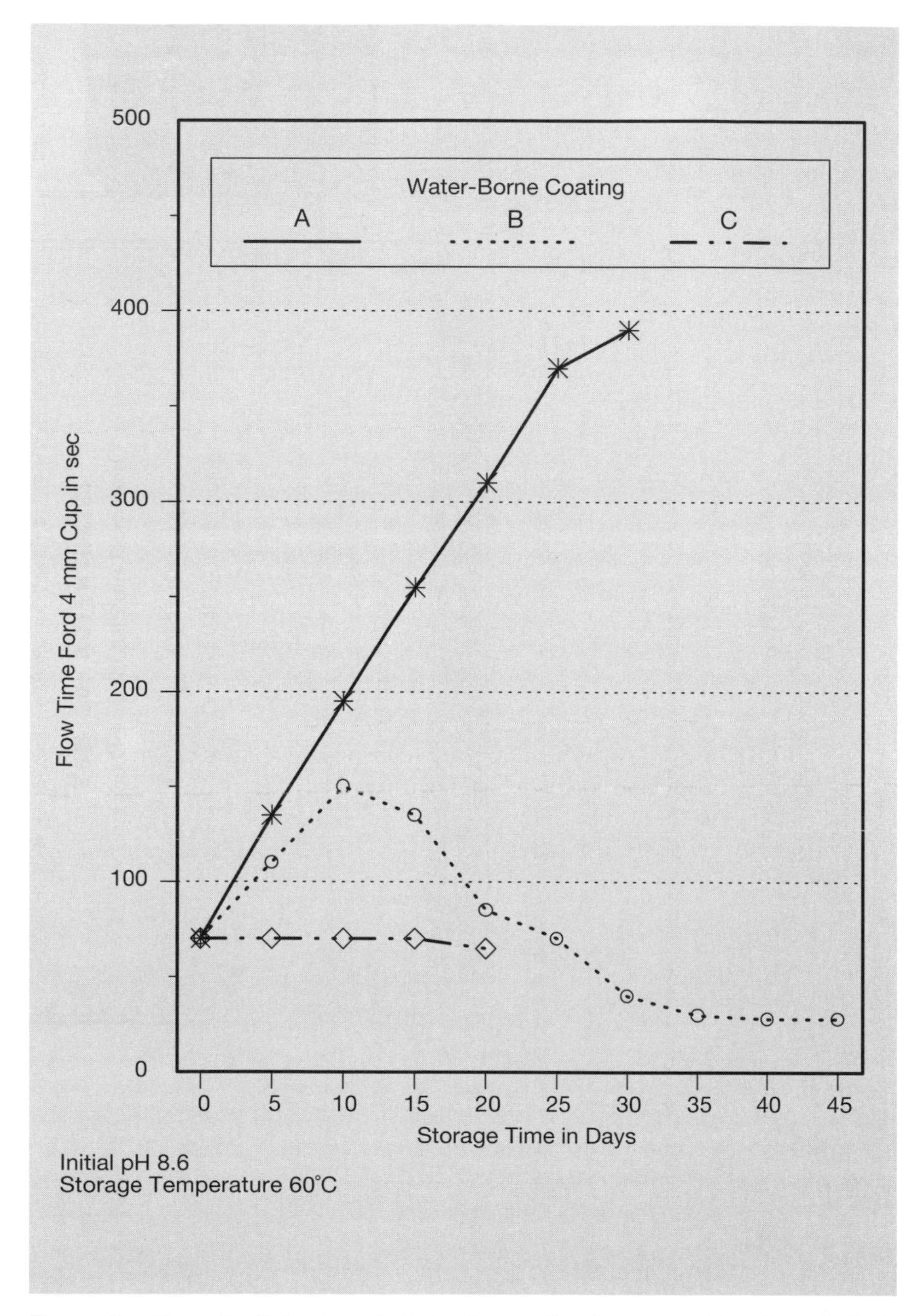

Figure 11 Viscosity Behavior of Water-Borne Coatings as a Function of Their Storage Time

10.4 Solubility and Viscosity Stability

Since a coating, as a rule is not applied immediately after production, but is stored for more or less long periods by the producer or applicator, the effects on change of the system with time must be known before production and must be considered during production. This is true especially for the coatings stability: Resin components must not precipitate, viscosity should remain as constant as possible or at least be adjustable later. Equally important occurrences such as pigment and filler wetting, constant flow and gloss, as well as constant reactivity will not be discussed at this point. The solution and viscosity stability of an aqueous coating depends on three factors:

- the nature of the resin
- the neutralization agent
- the auxiliary solvent.

If resins of differing chemical composition, for example of differing monomer construction, are studied, the following viscosity curves, as a function of storage time, are obtained (see figure 11):

- viscosity increases almost linearly
- viscosity increases and after passing through a maximum, drops again
- viscosity remains practically constant.

In all three cases, the pH drops from an original 8.6 to between 7.4 and 7.9. Investigations have shown that the acid number of the polyester solutions studied increases in parallel from about 49 mg KOH/g to 66 mg KOH/g. In all three cases, turbidity or even precipitation of the resin occurs after extended storage. To explain the results, a hydrolytic degradation of the polyester is assumed [42, 160]. The degradation occurs at the end of the resin molecule with splitting off of free carboxylic acid and a hydroxyl group-containing polyester. The formation of the carboxylic acid is the cause of the increase in acid number, the formation of the less hydrophilic OH-polyester is the cause for the separation of the resin from the solution. In the meantime however, it has become possible to, if not completely prevent this degradation process, to suppress it to an extent sufficient for industrial coatings purposes by selection of appropriate polyester monomers.

The storage stability of an aqueous coatings system is very definitely dependent on the amount of neutralizing agent used. Figure 12 shows the viscosity dependence, and figure 13 shows the pH dependence, on the storage time at 60°C and the amount of added amine. Accordingly it is useful to over-neutralize slightly to achieve better storage stability. A larger amine excess, to be sure, is to be avoided because of the acceleration of the polyester hydrolysis. This hydrolysis proceeds in parallel to the base strength of the amine (pKB values, Table 3).

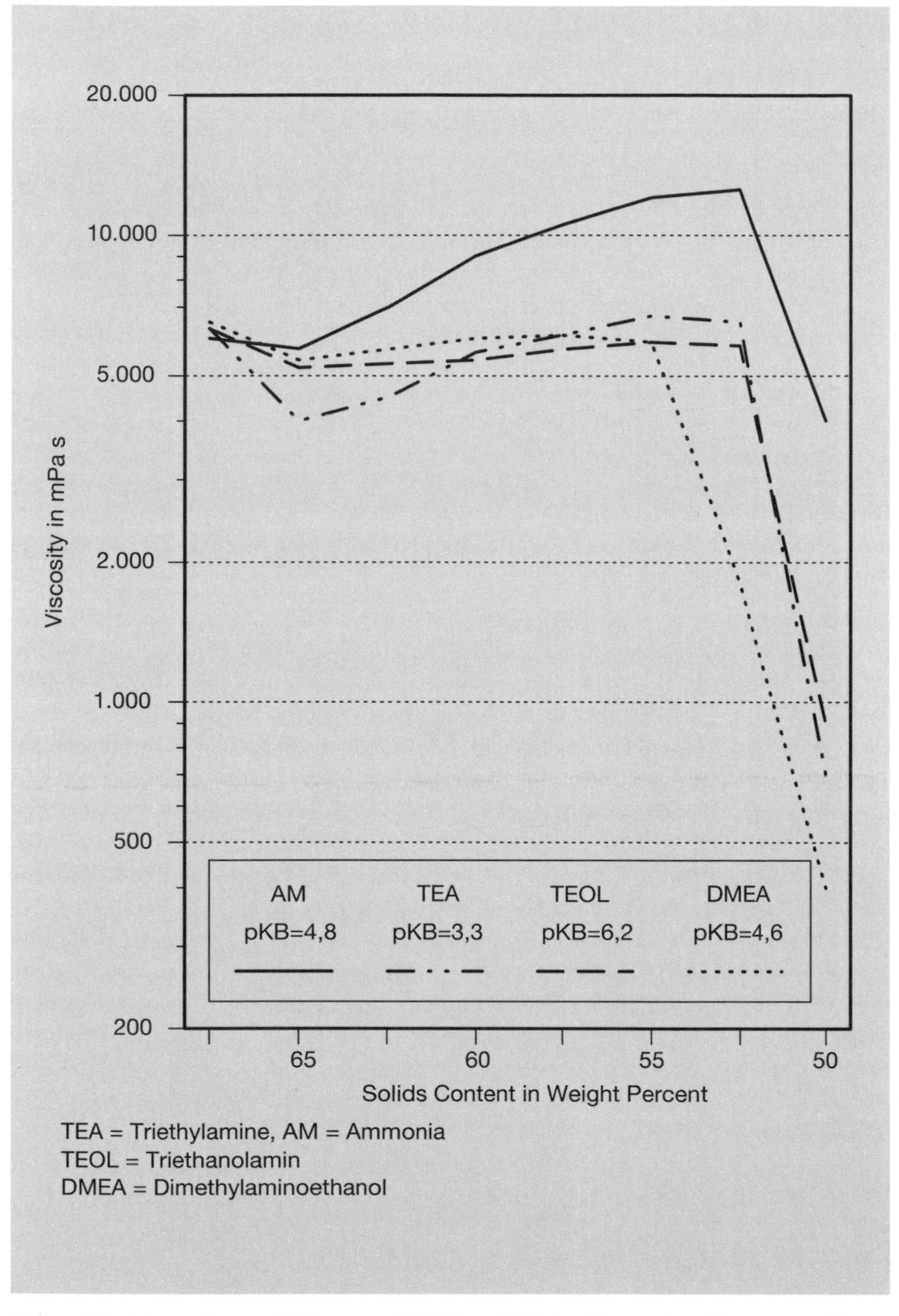

Figure 12 Viscosity and Storage Stability of Water-Borne Coatings as a Function of the Degree of Neutralization

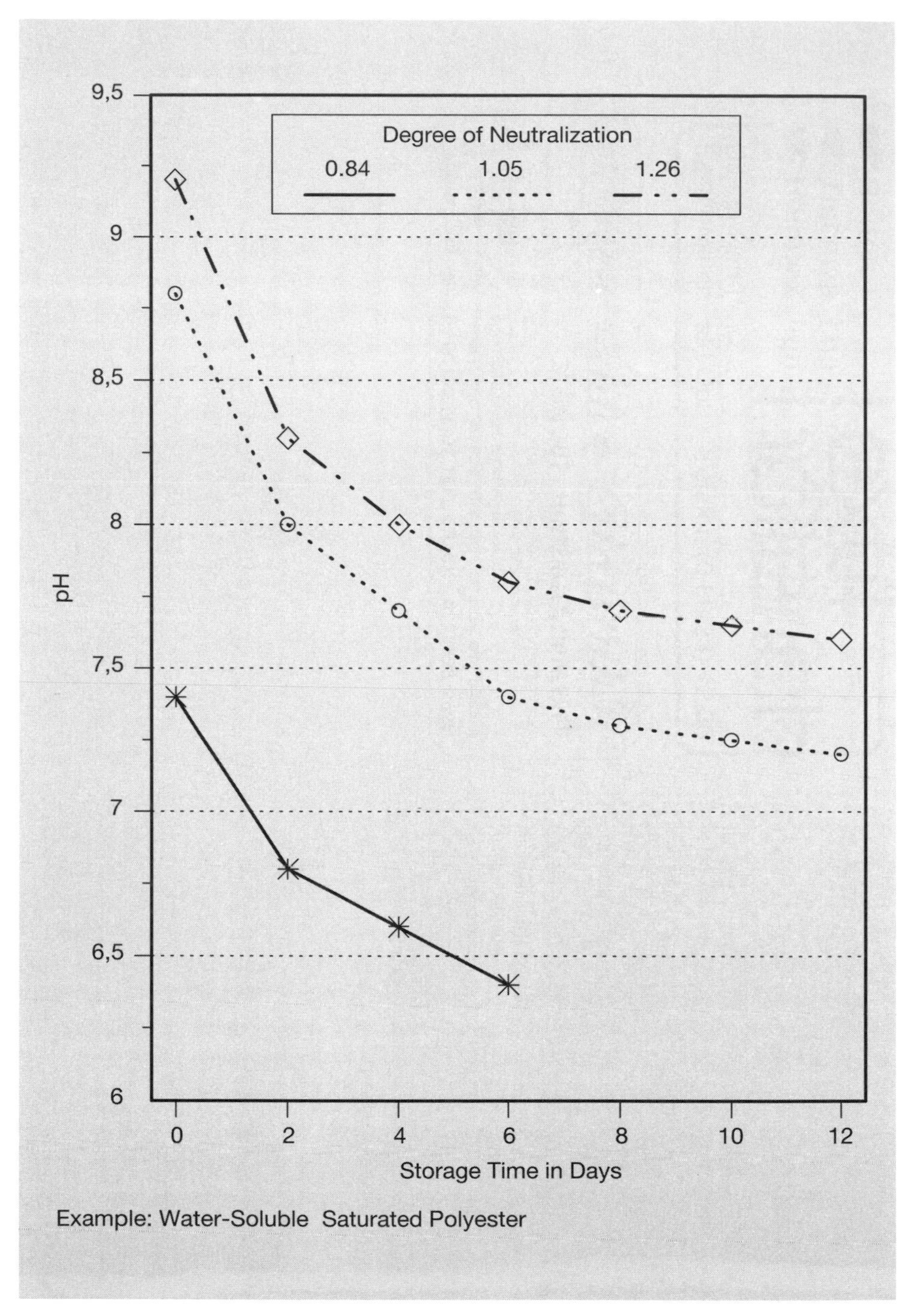

Figure 13 pH and Stability of Water-Borne Coatings as a Function of the Degree of Neutralization

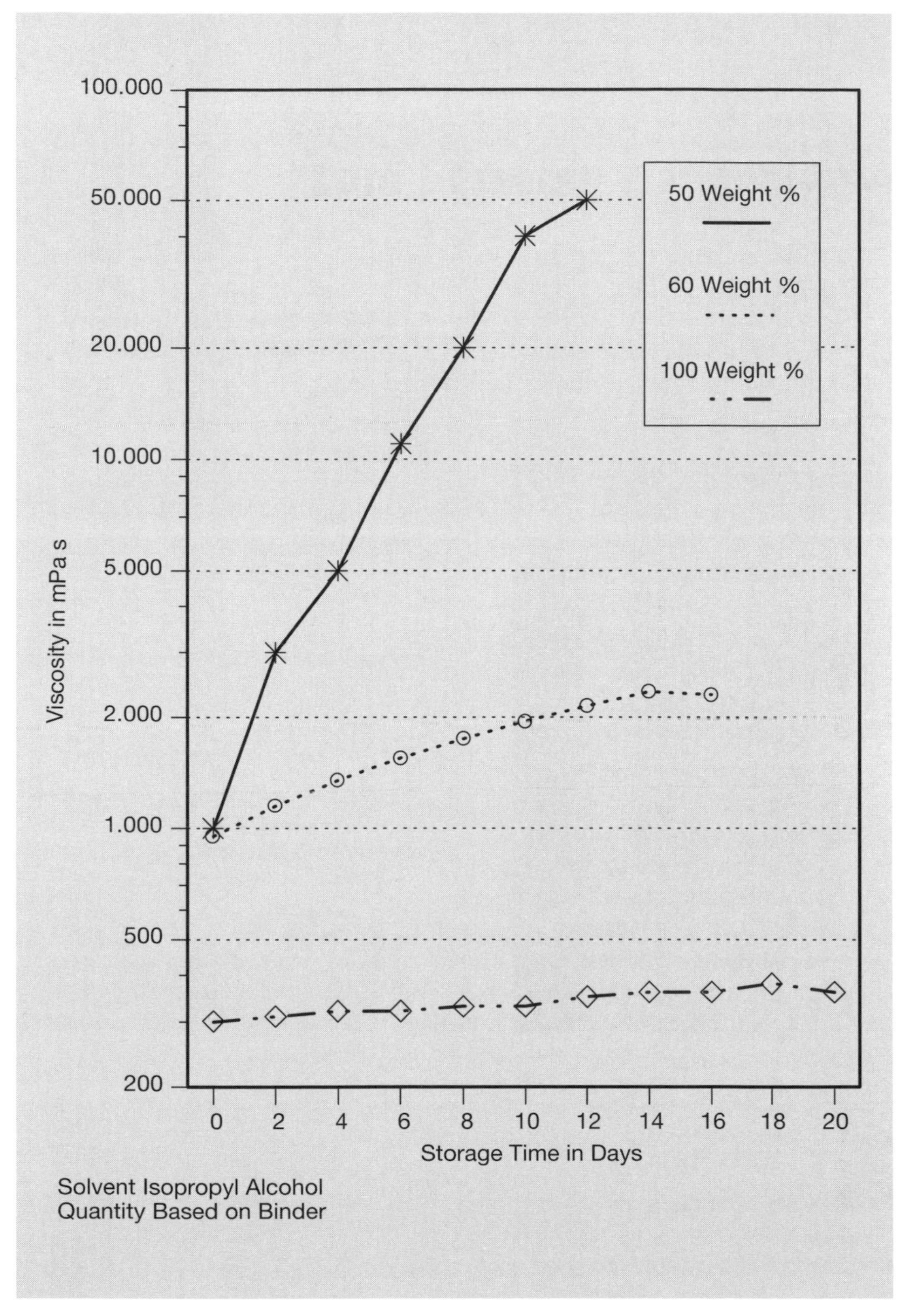

Figure 14 Effect of the Quantity of Solvent on the Storage Stability of Water-Borne Coatings

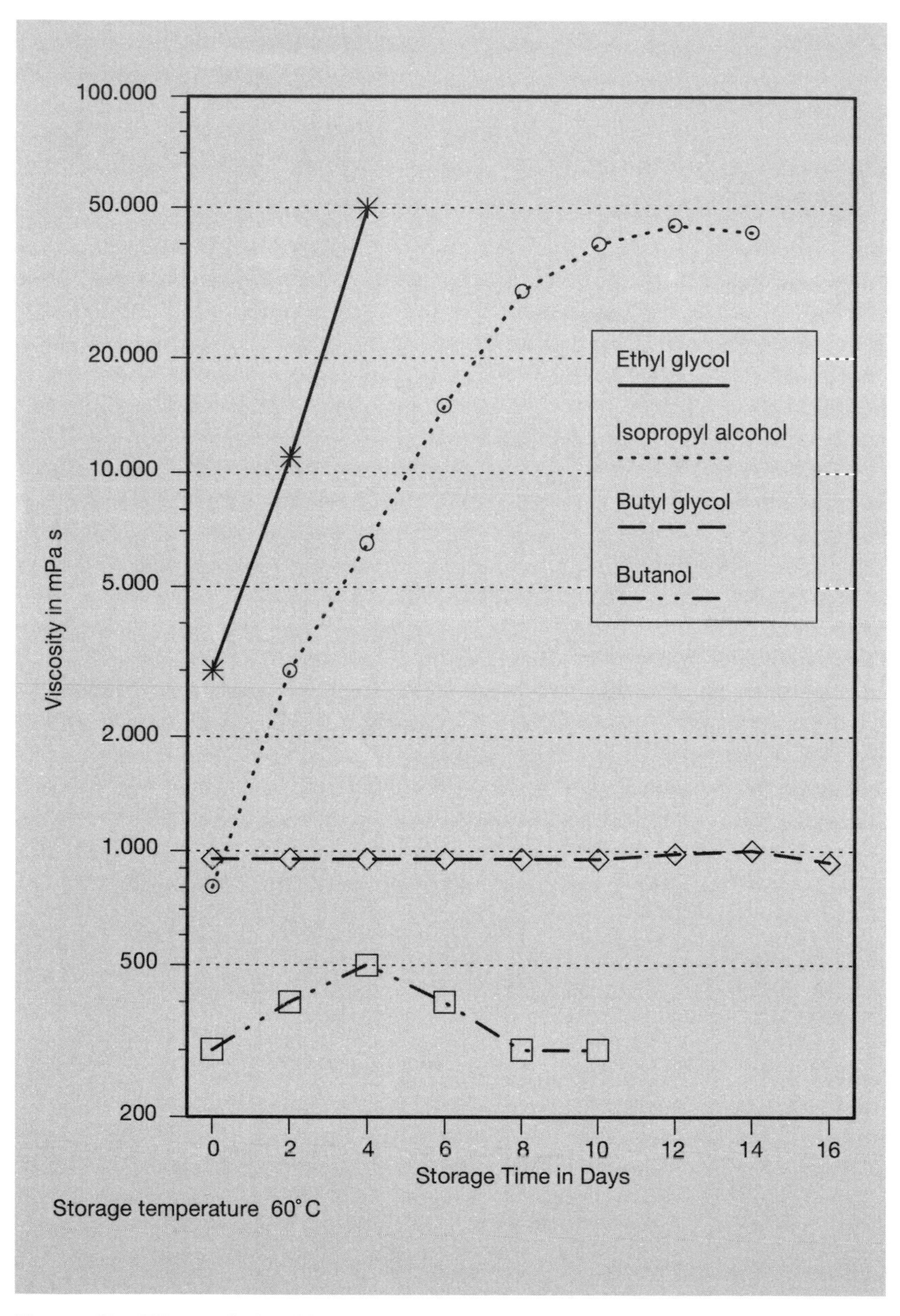

Figure 15 Effect of the Nature of the Neutralizing Agent on the Stability of Water-Borne Coatings

The separation of the polyester from the solution is greatly slowed if the amine also possesses an auxiliary solvent function. Here too, dimethylaminoethanol proves to be very effective.

The storage and viscosity stability of an aqueous coating are strongly dependent on the quantity and nature of the auxiliary solvent used. Those solvents, that have already been shown to be favorable in their viscosity behavior on diluting with water,are also especially effective for the improvement of the storage stability.

Table 13 Effect of the Neutralizing Agent on the Storage Stability of a Water-Borne Coating

Amine	pK_B Value	Binder Separation after Days	Degree of Neutralization after Precipitation
TEA	3.2	6	0.62
DMEA	4.7	11	0.76
Ammonia	4.8	8	0.84

Figure 15 shows the viscosity change as a function of the storage time at 60°C and as a function of the nature of the auxiliary solvent used. Ethyl glycol and isopropyl glycol are not realistic for use in the systems studied, since they cause a pronounced increase in viscosity in the coatings systems on aging. As before, butanol and butyl glycol are favored for attaining good storage stability.

The effect of the quantity of auxiliary solvent added is shown in figure 14. The viscosity behavior and resin separation from the solution on ageing can be significantly improved by the use of higher solvent concentrations in aqueous baking enamels. This is not the way to conceptualize coatings with reduced emissions, however.

A polyester-melamine system is not absolutely stable in aqueous-alcoholic media. Alkoxy melamine formaldehyde resins have a tendency to dealkylate even at room temperature, so that higher additions of alcohol are in a position to suppress the possible melamine resin self cross-linking (figure 16) [23, 35]. It has been further shown, that no reaction results in the aqueous phase at room temperature between the polyester and the melamine resin. The melamine resin self-condenses, however in dependence on the pH of the solution. At pH values between 7.7 and 8.2, no condensation can be established, while in the pH range of 6.5 to 6.8, the melamine resin cross-links with itself. From this it is evident that aqueous coatings should be adjusted to a medium pH range of 8.2 to 8.5. Higher pH ranges can result in degradation of the polyester.

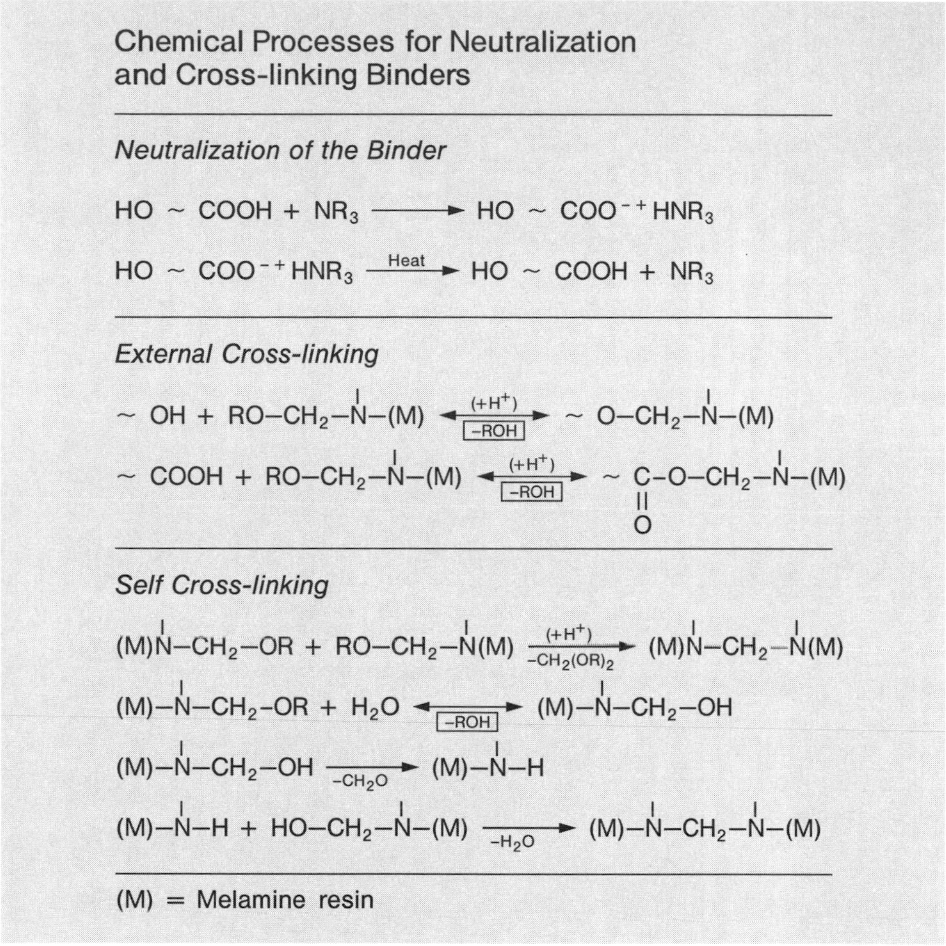

Chemical Processes for Neutralization and Cross-linking Binders

Neutralization of the Binder

$$HO \sim COOH + NR_3 \longrightarrow HO \sim COO^{-+}HNR_3$$

$$HO \sim COO^{-+}HNR_3 \xrightarrow{Heat} HO \sim COOH + NR_3$$

External Cross-linking

$$\sim OH + RO{-}CH_2{-}\overset{|}{N}{-}(M) \underset{\boxed{-ROH}}{\overset{(+H^+)}{\longleftrightarrow}} \sim O{-}CH_2{-}\overset{|}{N}{-}(M)$$

$$\sim COOH + RO{-}CH_2{-}\overset{|}{N}{-}(M) \underset{\boxed{-ROH}}{\overset{(+H^+)}{\longleftrightarrow}} \sim \underset{\underset{O}{\|}}{C}{-}O{-}CH_2{-}\overset{|}{N}{-}(M)$$

Self Cross-linking

$$(M)\overset{|}{N}{-}CH_2{-}OR + RO{-}CH_2{-}\overset{|}{N}(M) \xrightarrow[-CH_2(OR)_2]{(+H^+)} (M)\overset{|}{N}{-}CH_2{-}\overset{|}{N}(M)$$

$$(M){-}\overset{|}{N}{-}CH_2{-}OR + H_2O \underset{\boxed{-ROH}}{\longleftrightarrow} (M){-}\overset{|}{N}{-}CH_2{-}OH$$

$$(M){-}\overset{|}{N}{-}CH_2{-}OH \xrightarrow[-CH_2O]{} (M){-}\overset{|}{N}{-}H$$

$$(M){-}\overset{|}{N}{-}H + HO{-}CH_2{-}\overset{|}{N}{-}(M) \xrightarrow[-H_2O]{} (M){-}\overset{|}{N}{-}CH_2{-}\overset{|}{N}{-}(M)$$

(M) = Melamine resin

Figure 16 Cross-Linking Reactions in Aqueous Baking Enamels Based on Polyester/Melamine Resins

11 The Manufacture of Water-Borne Coatings

11.1 Introduction

While in the previous chapters the focus was more on theoretical concepts such as the chemical-physical structure of water-reducible binders, as well as the other formulation components, and their interplay in the end products, in the following chapters more of the practical side will be discussed. To this belong, in addition to the processing techniques and raw materials possibilities, also quality assurance and the manufacturing of water-borne coatings to be discussed in this chapter. In this connection, the raw materials to be charged for the production of paints and coatings will first be reconsidered. The procedures required for storage and handling, which are especially of importance for binders, will also be discussed. The manufacturing process and the customary mix and dispersion equipment are featured. It will become apparent that the systems encompassed by the key term "water-borne coating" require differing procedures. These arise for example from the differences between traditional emulsion paints with high PVC and glossy emulsion paints with low PVC as a result of differing consistency and differing quality demands. The same is valid for water-borne coatings in a narrower sense, depending on whether they are produced from water-soluble resins or with polymer dispersions, since both binder types behavior differently.

11.2 Raw Materials

The significant component, the binder should be at the forefront of consideration. Customarily, binders are judged based on their end properties and their applications behavior. Especially in water-reducible systems, however, paying attention also to the behavior in the production process is recommended. Aqueous binders are sometimes more problematic than classical solvent-containing systems. Especially critical are the following:

- the shear stability, that is, the resistance to mechanical effects (pumping, mixing),
- the chemical stability, e.g. toward salts, solvent materials (auxiliary solvents, film forming agents), but also towards the equipment used (tanks, piping, pumps, packing, filters, etc.),
- the resistance to bacterial attack.

The dispersing process is thus especially critical in as much as chemical, mechanical and heat effects all act together. As a tendency, water-soluble resins are quite similar to the conventional binders in their robustness, which can be explained by their closely related basic structure, while polymer dispersions often are more susceptible. Resin emulsions show more of an in between behavior. (see chapter 6).

The following points are to be noted particularly for the storing and handling of water-reducible binders:

- The storage temperatures must in all cases lie above the freezing point due to the poor freezing stability, so that storage out of doors is usefull only in exceptional cases. The temperature range between +5 and +30°C is favorable while +50°C should in no case be exceeded. Higher temperatures negatively affect the bacterial resistance of aqueous products (results: discoloration, gas evolution, odor). This theme was already dealt with in chapter 8.4, where regular processing cleanliness was especially recommenced [161]. To this belong the storage tanks (baked on binder, especially in upper areas), piping (blind angles) as well as mixing kettles and filling facilities. The use of preservatives is unavoidable with polymer dispersions. To check bacterial growth, a quantity above the minimal inhibition concentration is required: An insufficient dosage leads sooner or later to attack, an excessive dosage on the other hand is ecologically and economically unjustified. In addition to these biological effects, thickening can occur at higher temperatures in the case of water-reducible binders. These, such as skin formation (irreversible with dispersions!) also occur as a result of water evaporation from inadequately sealed containers.
- The effect of CO_2 from the atmosphere can cause a decrease in pH and thus instability. In certain cases, especially with polymer dispersions with high solids content, the inclusion of a stirrer to avoid creaming is recommended.
- The selection of materials for tanks and kettles must take into consideration the particular effects on aqueous resins [162]. Unalloyed steels should be coated (2-pack epoxy, phenolic resin coatings), highly alloyed non-rusting steels (only passivation of welding seams necessary), as well as fiber-reinforced polyester resins can be used as storage tanks without concern. In general, the formation of local elements must be avoided. Stainless steel or PVC or polyethylene are suited for piping. To minimize shear effects, the lines should not be too narrow and should not have kinked areas. Care must be taken with fittings that no components consist of aluminum or plated metals.
- Processing must be done as gently as possible. With good shear stability, centrifugal pumps can be used, otherwise membrane pumps are recommended. Here too, stainless steel or polypropylene are the materials of choice. Tubing pumps have proved successful in individual cases. In general, stainless steel filters should be used on the feed and discharge lines, whereby patent stoppers or quick-release filter screens are advantageous.

Water-reducible binders are not absolutely environment friendly, but should be considered as less damaging to the environment. Thus even classical polymer dispersions, that customarily contain neither solvents or amines, are considered as polluting water. Provisions are thus necessary to prevent their penetration into surface, ground and coastal waters. Included are dikes for storage tanks or other vessels (drums, containers). For rinse or other process water discharges, coagulation of the binders is realistic. The flocculated polymeric components can then be disposed of in a suitable land fill. One should be aware of the waste category of the binder or coagulate.

With respect to quality assurance, which is treated in depth in chapter 12, several recommendations for checking incoming binders will be given. To be named as generally valid tests are:

Determination of

- solids content
- viscosity
- surface tension
- pH

Evaluation of the film with regard to

- color
- turbidity
- gross particles
- cracks
- craters, etc. and

the determination of the screening residues.

To be added are for aqueous resins are

- the determination of the MFFT
- measuring the acid number (for a correct neutralization) or the
- Hydroxyl number.

The coatings manufacturer can be more confident about the raw materials supplier with a relevant certificate of analysis, which however does not obviate the necessity of incoming raw materials testing.

While the significant difference between solvent-containing and aqueous paints and coatings is based on the nature of the binder, no such general differentiation exists for pigments and fillers. Here principally, the same products are used as with classical coatings [104] (chapter 9). In the manufacturing process, the significant part of which encompasses the dispersion of the solid components in the binder as finely and as stable as possible, the tried and trusted criteria of the paint producer should be observed. Especially with the use of polymer dispersions, stabilization of the dispersed state of pigments and fillers must be supported by auxiliaries (chapter 8). With aqueous resins, as a rule, such dispersing agents are not necessary, occasionally however, wetting agents are used to achieve better gloss (chapter 8.6.). The customary pigments and fillers of today are usually pretreated so that the dispersion process is facilitated and good results are achievable even in a dissolver. For normal requirements, the use of pigment pastes has been shown to be advantageous, since dispersing is simplified and color tint variability is reduced. Especially with organic pigments, the effort is significantly higher and the management of the dispersion must be critically evaluated. For this dissolvers no longer suffice. The chemical activity of pigments and fillers must be observed in aqueous systems. Thus the pH value of an aqueous suspension and, above all in corrosion protection systems, the water-soluble constituents should be evaluated, since complications (stability of the binder, storage stability, corrosion protection properties) can arise. As incoming quality control, checking the particle size distribution, screen residues, oil number and, especially with pigments, the control of the color type, color strength and hiding power are recommended.

Additives and auxiliaries (chapter 8) are at least as important for the production, storage and application of water-reducible systems as for the quality of the applied end product. Thus the addition of neutralizing agents and auxiliary solvents is indispensable for the use of water-soluble resins if fully-formulated binders are not utilized (chapter 8 and 10). Anti-skinning agents are required to achieve sufficient storage stability for oxidatively drying resins. The necessity for preservatives in conjunction with processing cleanliness was already touched on several times (chapter 8.4 and 11). Water-reducible binders tend to foam with pumping procedures as well as in the dispersing process. This must be avoided as much as possible, since higher air content at filling (false charge quantity) and in the end product is defective. The admixture of air should be avoided or at least minimized during the mechanical operations mentioned to support the activity of the added defoamers (chapter 8.7.). Wetting and dispersing agents (chapter 8.6.) are indispensable to a stable dispersion of pigments and fillers. Otherwise viscosity changes (thickening, thinning) or settling can result, as can inadequate storage stability. In this respect, anti-settling agents are recommended for soluble resins. For dispersions, such auxiliaries usually do not suffice. True thickeners (chapter 8.5) must be used to establish viscosity and rheological additives must be used to achieve suitable applications properties. For incoming quality control, purely physical-chemical tests are often less predictive than practical applications evaluations, which show the effect of the product in a standard formulation.

11.3 The Manufacturing Process

It is impossible at this point to give a comprehensive picture of the manufacture of water-borne coatings. The paint and coatings industry includes a multitude of factories that extends from small manufacturers with local significance to globally-active firms. This structure results in differing sales strategies and product pallets. This fact must naturally also be considered in the course of production. A multifaceted product range with relatively small special products can only be mastered with a highly flexible strategy, which requires the constant change of raw materials in a facility or production train. The other extreme is mass production with automatic process control which at high throughput to be sure, allows for only limited flexibility. As a result, production can be more oriented to the *ab initio* method, which is based exclusively on the use of individual raw materials, or on a process that uses premixes (pastes) and thus allows more flexibility. In addition to economic and technical demands, ecological considerations are gaining in significance (i.e. more frequent product changes lead to increased cleaning and thus to more rinse water).

Water-reducible paints and coatings are not a uniform product group, but represent a large number of products of differing composition, on which widely differing demands are placed. This fact also must be satisfied by the manufacturing process. Thus the production process and the dispersion tools for a matt interior wall paint differ from those for a high-gloss automobile repair paint. As a result, only the significant principles of manufacturing can be discussed here. In general it is to be established that the production process for aqueous coating materials does not fundamentally differ from solvent-containing systems. Decisive however is the observation of certain basic rules which in their particulars lead to somewhat different process procedures than are known for classical paint systems.

The production process can be divided into several individual steps, in which the dispersion of the pigments and fillers into the liquid phase (water or binder/water) is the decisive procedure [126]. The individual steps are the following:

- Production of premixes
- Dispersion process
- Completion of mixing, let-down
- Testing, correcting or final adjustment
- Sieving, filling

The first step of the production process is the manufacture of a pre-mix for the dispersion process. A large part of the additives is already added here. Included are wetting and dispersing agents, anti-settling agents and thickeners, preservatives, defoamers, dryers if needed, corrosion inhibitors and film forming agents. In the case of aqueous resins, this is followed by neutralization, pH adjustment and the addition of auxiliary solvents before stirring in pigments and fillers [116]. The solids are added with slow stirring – beginning with the finest products – to favor de-gassing at relatively low viscosity and to minimize foaming. The composition of the pre-mix must be precisely adjusted for the demands of the dispersion process. Aqueous resin solutions can usually be used without problem in the dispersion process, since they are stable to shear and display good wetting properties. In the case of dispersions, this depends on the given binders and demands. For the customary dispersion paints that are prepared in a dissolver, the dispersion process is generally without problems. In the case of high-value coatings with high requirements for fineness of grind, color tint and gloss, a polymer disprsion can be overtaxed with respect to mechanical and thermal effects. The pre-mixing and the beginning of the dispersing are carried out in water alone with the addition of wetting and dispersing agents. By the application of corresponding shear forces, very good dispersion results are obtained, so the procedure can even be advantageous. The binder is only added in the let-down phase. Also the quantitative composition of the premix, i.e. the relationship of pigments/fillers to binder/water, and thus the viscosity must be selected so that the dispersion process proceeds optimally. As discussed below, the dispersion tools function especially well in a definite coatings consistency range. Further, the concentration differences between premix/dispersion and finished coating should not be too big. If substantial quantities of water, binder, auxiliary solvents and further additives must be added during let-down, the equilibrium established during dispersion can be disturbed (pigment shock, dilution shock) and in unfortunate cases, instabilities can arise. The actual quantity relationships thus represent a compromise between the demands of dispersing and let-down.

The function of the dispersion consists of three, closely related individual processes:

- dispersion of the pigments and fillers as fine as possible
- optimally wetting their surfaces and thereby
- attaining a stable state, which prevents renewed agglomeration.

The first step is not a matter of simply mixing solid and liquid components. To be sure, pigments and fillers consist initially of primary particles, which assemble in the dry state to so-called agglomerates. The function of dispersion is to vigorously crush these agglomerates, so that as much as possible the original primary particles are restored. Only thus can the effects of pigments and fillers such as coloration, coverage, gloss, protection (corrosion protection) thixotropy etc. develop completely and be utilized. The adhesive forces between the primary particles must be overcome to crush the aggregates. For this, energy must be expended which is introduced to the paint and coatings system in mechanical form by the dispersion tools. This is a matter essentially of pressure and shear exertions, which are conveyed by turbulent (e.g.dissolver) or laminar (three-roll mill) flow. Additionally, grinding media (e.g. bead mill) can be introduced. The individual apparatus types will be discussed in chapter 11.4. The fineness of the grind achieved is on the one hand dependent on the dispersion tools, but however is limited by the size of the primary particles. Crushing the primary particles is not possible with the machines used.

To be sure, cleavage of the agglomerates is only successful when the fine dispersion remains stable. This is achieved if the particles are completely enclosed by a layer of binder, additives and water so that renewed contact, and thus an agglomeration of the primary particles, is prevented. A prerequisite for such a shell is a certain affinity between the solids particle surface and the components mentioned above: Sufficient wetting must occur. This initially includes a displacement of the air adsorbed on the surface of the solid particles by binder and water, since air prevents intimate contact between the solid and liquid phases. This displacement process as well as the adsorption on the particle surface, is not only stabilizing. Both procedures also facilitate the dispersion itself, in that they promote dispersion of the agglomerates, since the adhesion of the primary particles among each other in the wet state is significantly less than in the dry state. For water-reducible binders, the wetting problem in general is relatively small as a result of the hydrophilic character of many fillers and pigments. Wetting agents, as are necessary with solvent-containing systems, can under some circumstances even be destructive. There are however certain differences with the aqueous binders due to chemical structure. The wetting mechanism is more favorable with water-soluble resins than with classical dispersions, since in the first approximation, they contain discrete resin molecules. These are decidedly smaller and more mobile than dispersion particles and thereby adsorb better on the particle surface and quasi imbed in the particle. This positive effect is used in the manufacturing process in that such shear-stable resins are specially introduced in the grinding phase, even if subsequently a dispersion serves as the actual binder (hybrid systems). After drying, the effect is again apparent, in that coatings based on water-soluble resins give more homogeneous and thus higher gloss films, than is possible with the dispersions currently offered in the market. Dispersing agents, which were discussed in chapter 8.6., must be used with the latter to obtain optimal pigment stabilization. They also attach themselves to the particle surface and prevent renewed agglomeration of the pigment particles through electrostatic repulsion. The optimal dosage must be experimentally determined for each formulation. The so-called rub-out test indicates whether a pigment dispersion has adequate stability and shows no flocculation. An unstable system results in color changes upon simple rubbing of the applied wet coating. This effect is attributable to the fact that agglomerations are significantly less stable in the wet state than when dry, so that they are destroyed by rubbing. Pigments and fillers that are not optimally dispersed also tend to settle.

It should be clear from the assertions made above concerning dispersion, that the selection of a procedure is strongly dependent on the coatings or paint system. A more or less intensive manufacturing process is required depending on the binder and paint consistency, the degree of gloss required and the pigments used. The function and capabilities of the individual dispersion tools will be discussed in chapter 11.4. To be noted also for example is that lamellar- or fiber-shaped fillers can be crushed by excessive mechanical forces and thereby be rendered worthless. This fact is taken into consideration by gentle mixing after the actual dispersing phase.

After the dispersing process begins the let-down phase. Here the remaining binders, water and other auxiliaries are added, so that the paint formulation is complete. The difficulty of concentration differences between the two steps was alluded to: A too crass change in concentrations can nullify the effect of the dispersion. Upon completing the grinding stock an additional charge of defoamer is frequently advantageous. Defoamers lose a part of their activity in the dispersion process and must be refreshed for the subsequent production steps (stirring, screening, filling off) as well as for later application. Ideal, if also equipment intensive, is degassing of the mixture in a vacuum mixer or in general the manufacture in a vacuum disolver. Additional auxiliaries, which are added in the end phase, are anti-skinning agents for oxidatively-drying binders, rheological additives, auxiliaries, e.g. for the control of drying, and film forming agents. A precipitous treatment with such agents can lead to incompatibility and coagulation of polymer dispersions. This must thus be done carefully and, if necessary, after dilution with water.

The penultimate step of coatings manufacture is release testing and if needed, correction of the coatings composition and properties. In the sense of quality assurance [163] and finally also out of purely economic considerations, it is unfavorable to carry manufacturing through to the end and then to correct any possible shortcomings. More effective is the incorporation of intermediate tests for the early recognition and elimination of mistakes. Such a procedure is more difficult with the above mentioned *ab initio* methods than when working with pre-mixes and pastes, which have already been subjected to testing. It can also be advantageous to establish initial formulations loosely, and then only after testing, to make the fine adjustments. This is not a matter of eliminating mistakes, but a targeted final adjustment. In general, it must be noted in the release testing, that many properties are not finalized until after a certain maturation.

The final testing naturally relates predominately to the specifications properties, but frequently technical applications tests are also included. These are determined by the applications area of the coating and also the profile of requirements that the customer presents. The following are given as standard tests:

- Density control (especially important for volumetric filling off)
- Determining the solids content
- Viscosity measurement, which can range from a determination of the flow time to rheological investigations of the coatings, and
- pH determination.

Achievement of the required

- fineness of grind is checked after the dispersing procedure.

Before filling off, in each case the

- screen residue should be determined to prevent screen or pump blockages.

Bacterial attack represents a constant danger to aqueous systems:

- The bacterial count must thus be regularly checked.

In general colored paints must always be

- investigated for color tint as well as
- degree of gloss.

Classical interior dispersion paints represent quite price worthy products, but also they must fulfill certain quality requirements with respect to

- coverage
- degree of whiteness
- matting
- scrub resistance [81].

Further properties, depending on the application, are

- hardness
- elasticity and
- adhesion,

which are controlled by diverse tests such as pendulum hardness or Buchholz hardness determinations, impact and cross hatch tests. With corrosion protection products, the usual resistance tests (salt spray, humidity resistance, exchange tests) are carried out and the tendency to form flash rust is checked. In the finalized test program, it must be considered which tests are always part of the release testing. In this decision, customer requirements are of very important significance; the tests should however generally be quick to execute, predictive and cost effective.

After release of the material, filling off can follow, in which in each case a screening is to be introduced. Metal and synthetic sieves are used for screening, in special cases oscillatory sieves simplify the process. Here too, the shear stability of the coating system must be considered. Depending on need, cans, drums or containers are used as packaging units. To be noted, as also in the production facility, is the selection of the correct material. The PE containers used for architectural paints are without problems. In metal packaging, as especially customary in the paint realm, a suitable and complete interior coating is required. If the coating is lacking, rust formation can occur, especially on the lid. In keeping with the newest determinations regarding packaging materials, the trend to large containers, that ideally cycle between producer and customer, especially with large users, will increase if technically feasible. Here obviously, careful cleaning of the containers must be ensured. Fire is not to be feared during storage of aqueous products due to the small solvent content, but freedom from freezing if necessary. To be sure the freezing point can be lowered by the addition of certain solvents (alcohols, glycols, etc.), but this hardly makes sense with environmentally-friendly products. Also such systems that do not completely coagulate on freezing and again thaw, are changed by such a procedure (viscosity increase, spotting), so that their use is restricted. For security, it is recommended therefor that storage temperatures not be lower than +5 °C. In general in this regard it must be remembered that also aqueous products are not completely innocuous ecologically and that pollution of the ground water during storage must be prevented.

In conclusion of this chapter the already mentioned key word cleaning will be discussed. Especially with dispersion systems, the coatings producer must exercise care. Dispersions form films irreversibly and are thus insoluble in the "solvent" water, and to be sure already in a relatively early stage of the film formation process. Also in organic solvents, there is no solubility of dispersion film, but at best an ability to swell, which makes possible mechanical removal. Since the latter is tedious and especially connected with the undesired introduction of solvents, film formation must be avoided at all costs during the entire paint manufacturing process. In this aspect, it can be favorable to add the binder late, e.g. during the let-down phase. Film formation can be prevented by working in a closed system. Paint residues are then removed by thorough rinsing, whereby however for ecological reasons rinse water must be kept as small as possible. Here too it is necessary to find a reasonable compromise between technical, economic and ecological demands. Technology and economics collide with each other if aqueous and solvent-containing products are to be co-produced in a facility. If these differing systems interact, in addition to undesired mixing, coagulation of the water-borne coating can easily occur and thus blockages of piping, screens, pumps, etc. As a result, strict separation of both product groups is recommended.

11.4 Dispersion Equipment [13]

Dissolvers are among the most widely used tools [164]. They are characterized by high capability with simple construction, by which *inter alia* there are good cleaning possibilities. Several dissolver types can be differentiated. The basic model consists only of an axle with a dissolver disc. In addition there are double wave dissolvers with three offset dissolver discs. Further a mixing kettle can also be equipped with an eccentric dissolver and a stripper. While normally a dissolver disperses and maintains the material flow, these functions are largely separated in such an arrangement. A further variation is the so-called vacuum dissolver, in which the incorporation of air into the coatings material is optimally prevented. To support the dispersion effect, the dissolver discs have teeth, points and bars which work as grinding media. The actual dispersion is achieved by the fact that the dissolver creates a turbulent flow, by which the solids agglomerates exchange between high and low pressure in zones and are thereby divided (shear forces).

In addition to the wetting which is in progress, the solid particles also act reciprocally as grinding media. From this it is understandable that optimal flow relationships must be present in the dissolver to achieve good dispersion. Prerequisite are establishing the correct viscosity of the material to be dispersed and observing the necessary geometry [81]. If the viscosity is too high, the material will not be properly treated. Frequently the paint in the immediate area of the disc is soft as a result of heat evolution and flows, while the rest of the material remains immobile: The dissolver stirs in a hole in the material without dispersing. If the viscosity is too low, no calm flow results, energy transfer is not possible and the dispersion effect does not occur. The formation of the so-called doughnut effect, whereby a small portion of the dissolver disc remains visible when viewed from above, represents the optimal flow pattern. For this, the establishment of the optimal bulk viscosity and geometry are as indispensable as the selection of the correct fill depth.

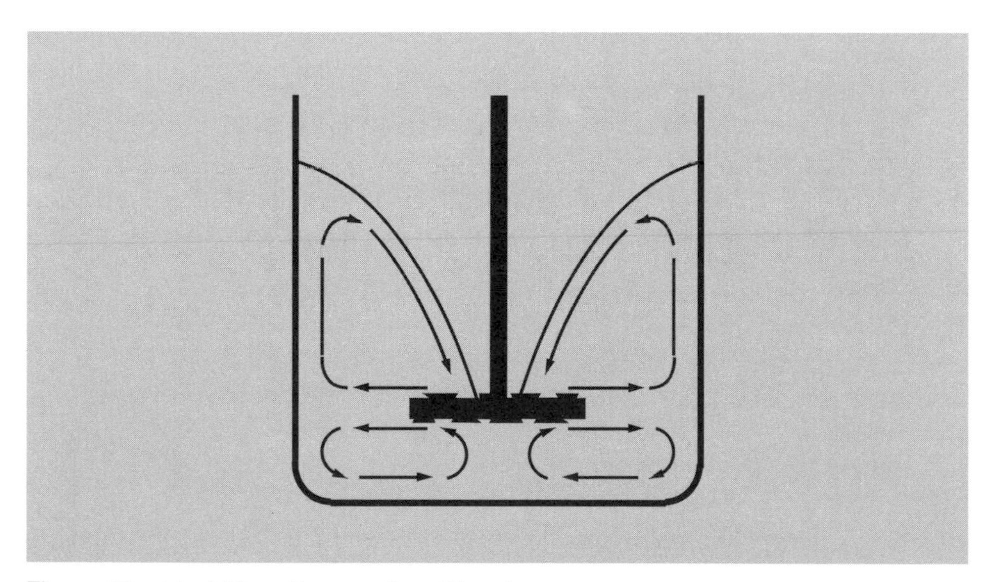

Figure 17 Ideal Flow Process in a Dissolver

The rate of revolution of the disc is decisive with respect to the energy transferred by the dissolver. With the normal dispersion paints, 10 m/sec suffices, with true paints higher values are necessary. The temperature increase with dispersion is desired, especially since it makes clear that energy is being transferred to the material being dispersed. With aqueous products, never the less, the shear and heat stability must be noted: a temperature of 50°C should not be exceeded. Here too it can be advantageous to disperse an aqueous color paste and subsequently to let down with the binder. The capability of dissolvers is limited. They are suited above all for dispersing easily-wetted pigments of medium fineness. For very fine pigments, as for example organic color pigments, their dispersion capacity is insufficient. In general, after a dispersing time of 20 minutes, a further increase in the dispersion effect is no longer possible, so that other processes must be selected when the results are not satisfactory.

A totally different type of dispersion tool is represented by the roller mill, in which above all highly viscous media, such as color pastes and printing inks are dispersed, especially where pigments that are difficult to wet are concerned. In the realm of water-reducible products, this tool is not widely distributed.

Today almost only three-roll mills have larger significance, while one-roll systems (pebble mills) are used more seldom for the production of pigment pastes (medium fineness) for dispersion paints. The dispersion energy introduced into the system results from the shear gradients between the rolls rotating at different speeds, the viscosity of the medium and the space between the rolls. To be noted is that the material not contain coarse agglomerates that could damage the surface of the rolls. The PVC of the mixture also has an effect on the dispersion effect. With increasing pigment content, a reciprocal rubbing of the agglomerates occurs, which in addition to the shear forces furthers the dispersion. A disadvantage is the relatively low throughput of roller mills. The one-roll mills can also be used for screening, since the large particles gravitate to the top where they can be removed.

Ball mills belong to a third type of dispersion equipment. Here balls kept in motion by rotating tumblers, transfer their energy in the form of pressure and rubbing to the pigment agglomerates, thereby producing similar conditions in the coatings films as with roller mills. Indeed the consistency must be selected so that movement of the balls is not hindered. In general it can be noted that ball mills today no longer have the significance of earlier years. This tool however produces a good dispersing result, is not problematical in use and requires no special attention. However it also has the following disadvantages: Relatively low useful volume (ratio of balls to material to be ground of between 1:1 and 2:1; ca. 75% fill volume) and as a result of exclusive batch processing, only low throughput, long dispersing times, difficult cleaning, i.e. low flexibility, and very high noise levels while in operation.

An improved range of properties is displayed by *attrition mills*, which function according to a similar principle as ball mills. The grinding medium which is also used here is however set into motion by the agitator discs and is also accelerated. Formerly Ottawa sand was especially used as the grinding medium, today on the other hand, glass, ceramic (e.g. Al_2O_3, ZrO_2) and steel balls are used. The finer the medium, the higher the effectiveness, the more tedious the sieving process that again separates grinding medium and material being ground. Today balls as small as 0.2 mm are available. The significant advantage of attrition mills lies in the possibility of continuous processing which permits higher throughput. The capacity with respect to the dispersing process is very high, so that organic pigments can also be dispersed. By controlling the rate of throughput, the dispersion results can be influenced. In addition to the usually vertically oriented tools, today more user-friendly, faster horizontal mills are increasingly used. With aqueous products the use of closed equipment is recommended to prevent irreversible film formation. Attrition mills are still in a phase of intensive further development, to achieve still shorter processing times with improved dispersion [167]. The so-called Turbomill should be mentioned here which can be described as a combination of dissolver and attrition mill [168]. The arrangement fundamentally corresponds to a dissolver, where however the dispersion disc is replaced by a steel screen basket, in which grinding media are kept in motion by an agitator. The material is treated in the kettle, similar as with a dissolver, and dispersed to high fineness. The advantage compared with classical attrition mills, is in the higher flexibility and faster purification. In general, manufacturing using pastes is advantageous with high capacity attrition mills, so that the shear-sensitive binder is introduced only in the let-down phase.

12 Quality Assurance of Water-Borne Coatings

12.1 Introduction of a Quality Assurance System

The interest in quality and quality assurance has increased worldwide in recent years. Today enterprises place special value on the high quality of their products and services to satisfy the increased expectations of their customers and thus be able to successfully market their products. As a rule, enterprises also set high goals for the quality of their products in the past, but new developments in the marketplace demand new forms for realizing, maintaining and documenting quality.

Numerous enterprises invest substantial means to install and expand quality assurance (QA) systems. Thus, the management of the enterprise works out a strategy and goals for quality. Since numerous influences on the special product palettes of the enterprise have an impact, and the quality assurance system is determined by both the large and individual goals, there can be no universally-valid quality assurance system. In the standard ISO 9004 (EN 29004) however, numerous important elements for the installation of such quality assurance systems are described. In ISO 9000 (EN 29000) is given a schematic for the selection and application of the quality management and quality assurance standard ISO 9001 to 9003 (EN 29001 to 29003), in which three models for the external documentation of quality assurance are described. The types of documentation are defined to a varying extent.

ISO 9001 should be used, when a firm guarantees the fulfillment of established requirements relative to numerous phases. The phases can include development, production, assembly and customer service. ISO 9002, on the other hand, should be used when the fulfillment of requirements relating to production are to be warranted, while according to ISO 9003, only the fulfillment of established requirements relative to final testing is guaranteed.

12.2 Advantages of a Quality Assurance System

A major advantage to enterprises for the introduction of a QA system based on the above norms is certification by an independent agency. This certification confirms the quality capability to the customer of the certified firm so that the customer can dispense with his own time-consuming audits. The DQS (German Society for Certification of Quality Assurance Systems, Inc) is a nationally and internationally recognized, independent agency, which has created rules for the execution of quality audits, for the certification of auditors and for the use of the standards ISO 9001 to 9003 as foundations for quality audits and certificates [169].

A certificate about a QA system must not be confused with a product certificate. The first confirms the general quality capability of the supplier, the product certificate refers to the quality of specific products at a specific time.

An additional advantage of a QA system to a supplier is represented by trustworthiness on the part of the customer, for the supplier cheats a framework with the QA system with which the requirements of the customer are fulfilled to his complete satisfaction and can also be verified. With this quality standard, the supplier can become the favored partner of the customer and thus overcome the competition. Today it is no longer the exception that customers conduct a supplier evaluation and the product thereafter is only obtained from the so-called A suppliers, which received the most points in the grading.

Finally, an increase in profits is tied to better and documented quality. An investment must however, be made for the introduction of quality assurance, especially in the area of error prevention in the manufacture of products. When these measures take hold, the costs for complaints and for final testing decrease. When the risk of producing a defective product is definitely lower, the analytical expenditure can also be correspondingly reduced. The nucleus of quality assurance is thus not the final testing, but quantitatively guaranteeing the process capability. Practice already shows examples of an increase in profits on introducing and establishing a QA System [170].

12.3 Quality Assurance for Manufacturers of Water-Borne Coatings

In this chapter the quality assurance of water-borne coatings will be discussed, and especially the testing of raw materials, manufacturing and the final testing of the water-borne coating. For this problem, the use of ISO 9002 is the most suited. Thus the most important points of this norm will be discussed here.

12.3.1 Organization

According to ISO 9002, the responsibilities and assignments of employees must be established unambiguously. The organizational independence of the employees who establish quality deficiencies of the products by testing and who initiate measures for improvement must be observed. The leadership of the enterprise must verify the fulfillment of the standards at regular intervals by internal audits. The QA system must be documented to be in agreement with the standard through procedures and reports and must be correspondingly realized in practice. As mentioned previously, there is no generally valid standardized QA system, thus each enterprise must interpret the requirements of the standard and relate these to the specific problems of the enterprise.

A quality handbook and quality assurance plans that amplify the relevant requirements in more detail should be established. These documents must be approved by authorized personnel. Care must taken that the documents are distributed to all position, that are authorized to contribute to the functioning of the QA system. The distribution, altering or removal of expired documents must proceed in a controlled manner.

12.3.2 Procurement of Raw Materials

The producer of water-borne coatings must procure the corresponding raw materials for their production from suppliers, for in only a few cases will the producer of water-borne coatings produce a portion of the raw materials internally. These suppliers must be carefully located by the manufacturer. Purchasing documents must be created, that describe the type of product and which contain the specifications and test methods and other technical details. The manufacturer of the water-borne coating can request a certificate from his suppliers. In the German Standard DIN 50 049, various types of certifications are described.

In a *Certificate of Compliance with the order,* the producer confirms, in the form of a text, that the delivery is in agreement with the order acceptance. This certificate contains no test results. The specifications of the product information can serve as a basis for agreement for acceptance of the order.

In a *Test Report,* the producer confirms that the delivery corresponds to the agreement of the order acceptance by test results from the continuing testing of the delivered product. For a test report, the tests need not necessarily be done directly on the material delivered, in contrast to the Manufacturer's Test Certificate. The *Manufacturer's Test Certificate* contains test results from the delivered goods.

In an *Inspection Certificate,* the test results for the materials delivered are confirmed by experts that are independent of the producing concern. All these types of certificates however, do not absolve the manufacturer of the water-borne coating from the responsibility of controlling incoming goods.

12.3.3 Testing of Raw Materials

As an example an enterprise will be assumed that produces a water-borne coating based on a saturated polyester as binder. This polyester dissolved in butyl glycol is produced by a supplier who issues a product information bulletin with the following specifications (see next page please):

The enterprise can now request a Manufacturer's Test Certificate with the precise properties of each shipment. The values must all lie within the boundaries given in the specifications. In addition,an incoming test of the most important properties should be conducted.

Table 14 Specifications of a Saturated Polyester

Property	Value	Unit	Test Method
Baking Residue	82–84	Weight %	DIN 53216
Acid Number	42–48	mg KOH/g	DIN 53402
Hydroxyl Number	89–90	mg KOH/g	DIN 53240
Viscosity at 23°C, 60% Solids	500–800	mPas	DIN 53018

To produce a coating, the polyester must be neutralized with an amine. To correctly adjust the degree of neutralization, an accurate knowledge of the acid number is of importance. Therefore, this should absolutely be taken up in the incoming testing. Knowledge of the hydroxyl number is decisive for the ratio of binder to cross-linker in the formulation (in this example: modified water-reducible melamine-formaldehyde resin). The viscosity is an indirect measure for the molecular weight of the polyester, which influences the technical applications properties of the water-borne coating, as e.g. the elasticity. Therefor the latter two properties can be taken up in the control of incoming goods.

To neutralize the acid groups, dimethylaminoethanol or dimethylaminomethylpropanol can be chosen, for example. These amino alcohols impart better reducibility with water and better storage stability, since in addition to salt formation, they function as auxiliary solvents (see chapter 8.1 and 10). When purchasing these raw materials a certificate can similarly be requested. As incoming control, a purity determination by a GC analysis or also by the refractive index or density can be conducted.

With water-borne coatings, difficulties in surface quality must always be anticipated. Frequently there are problems with flow, drying and gloss. In contrast with solvent-containing coatings, in which mixtures of solvents are used, no temporally extended smooth evaporation can result (see chapter 5 and 14). In aqueous systems, pigment wetting and dispersing are also more difficult than with solvent-based coatings (see chapter 9). These properties can be improved by incorporating auxiliary solvents and additives (see chapter 8).

Butyl glycol functions as solubility promoter in this example, in which the saturated polyester is present as an 83% solution. If a solid resin is obtained from the supplier, an auxiliary solvent must also be purchased. Water miscible solvents with extended alkyl chains should be chosen for this, since they effect a favorable dilution behavior (low viscosity) (see chapter 10). Butanol or butyl glycol function as such auxiliary solvents. Here too, a product identification by density or refractive index and a purity determination by gas chromatographic analysis can be done.

If the manufacturer would like to offer a white baking enamel, titanium dioxide is used as a rule. Most commercial titanium dioxide pigments are surface treated inorganically or organically to improve the technical applications properties. For example, a pigment treated with aluminum oxide with an isoelectric point of 8 to 9 can be used for the water-borne coating. This has the advantage, that at a pH of about 8, the neutralized system will not form agglomerates. Typical specifications or characteristic data for such pigments are the titanium dioxide content, the content of additional constituents, density, refractive index, sieve residue, oil number and the relative dispersity. The water-borne coatings producer should also request a certificate from the pigment producer. As an additional incoming control, a check of the purity by density and refractive index should suffice.

A modified melamine resin, which is present either dissolved in water or which becomes water-reducible after mixing with the binder and neutralization, is used as the cross-linking component. A certificate could contain the non-volatile components, viscosity, Gardner color number, density, flash point or the equivalent weight. The latter gives the quantity of cross-linker in grams that react with a mole of hydroxyl, carboxyl or amide groups. As a rule, a purity test is carried out as incoming testing.

A water-borne coating produced on a laboratory scale, which is applied on a substrate and subsequently subjected to technical applications testing for which internal specifications exist, can also serve as an incoming control for several of these raw materials. These can for example be hardness, elasticity or adhesion tests and an evaluation of the flow and the surface properties. This is especially effective when the producer of a water-borne coating has garnered good experience with the quality of the raw materials over years and would thus like to spare a large part of the incoming testing. If one property falls outside the accepted range, each individual raw material must be tested.

12.3.4 Manufacture of the Water-Borne Coating

According to ISO 9002, the manufacturer of the water-borne coating must demonstrate the capability of steering the processes in the production. For this steering, measures must be planned and defined. These are described in the form of operating instructions, which are viewed as quality documents. In these operating instructions, the progress of the individual steps and the intervention conditions are enumerated. In addition, it must be established that only suitable production facilities and reasonable working conditions are used. During the production process, supervision of the relevant process or product criteria must be carried out.

Each criterium is subject to statistical variation, which are however relatively small and are attributable to factors that cannot be influenced. During the process, these variations can increase and exceed a previously established value, so that proper controllable factors affect the process. The consequence arising is an intervention in the process. The use of control charts after Stewart [171] supports the required control of the process. In these control charts, the test values are entered in chronological sequence. The control chart contains warning and intervention limits, which can be derived from a statistical computation of an extended observation of the undisturbed process. If a value lies outside the intervention limit, one can assume that a controllable factor is out of control. A value outside the warning limit, but within the intervention limit, should awaken attention and lead to extraordinary sampling. Control charts serve as quality records and should be preserved in an appropriate way.

The use of these control charts is certainly not always realistic for a water-borne coatings producer, since this is not a production process in the real sense, but instead the manufacture of the water-borne coating takes place in several discrete steps. Frequently a single manufacturing step is not controllable in the above described manner. It is for example, not feasible to follow with the help of control charts, the process of dilution with water, after completion of the solution of the polyester in organic solvent and neutralization. Instead the calculated quantity of water is added, and subsequently by a test it is determined whether the solids content has been adjusted as desired.

A realistic control of a value during a process could result from the manufacture of an aqueous dispersion, namely the testing of the particle size of the binder upon dispersion in water (see also chapter 11.3.).

12.3.5 Final Testing of the Water-Borne Coating

It must be established that all final testing is conducted according to the quality assurance plan in keeping with the quality assurance operating instructions. The water-borne coating must not be shipped before all tests are completed and found to be in order. For this, suitable analytical reagents with the required accuracy must be prepared. The analytical reagents must be supervised, calibrated, justified and maintained, and this procedure must be documented.

A product that does not meet established specifications, must be barred from sale and be protected against erroneous shipping, e.g. by conspicuous labels. It must be established who assumes the responsibility for disposition of defective products. This position then decides, if a re-work, a reuse for other purposes or scrapping will be done. The re-worked products must naturally be subjected to renewed testing.

Quality assurance however goes a step further and analyzes the causes of failure, to avoid repetitions. Thereupon, error prevention measures are developed, introduced and their effectiveness is supervised. The goal of quality assurance, after all, is not to provide perfect all-encompassing final testing, but to constantly produce products with no quality shortcomings, so that final testing and costs can be minimized over time.

As an example of final testing, the polyester described in chapter 12.3.3 will be chosen here, which can be used e.g. as binder for a water-soluble spray paint for windshield wiper holders. In the manufacturing process of the coating, the polyester was neutralized with dimethylaminoethanol and a melamine resin, zinc phosphate and various additives were added.

The coating is tested according to the testing plan in observance of the operating instructions. The values must lie within the internal and external specifications. The following determinations are carried out directly on the paint and can serve as external specifications:

Table 15 Final Testing of the Water-Borne Coating

Property	Value	Unit	Test Method
Viscosity, DIN-4-Cup	180–200	s	DIN 53 211
Flash Point	100	°C	DIN 51 758
pH	8.2–8.8		

The coating is then diluted with water to a working viscosity of 50 s (DIN cup 4) and sprayed on Bonderite panels. Subsequently it is baked for 10 minutes at 180°C. The dry film thickness should be 20 μm. The following technical final tests can serve as external or internal specifications:

Table 16 Industrial Testing of the Water-Borne Coating

Property	Value	Unit	Test Method
Pendulum Hardness	125–135	s	DIN 53 157
Square Cut Adhesion	0	–	DIN 53 151
Cupping	6.7–7.3	mm	ISO 1520
Reflectometer Value (60°)	35–40	%	DIN 67 530

The tests are carried out under standard conditions according to the German Standard DIN 50 014. After testing is completed, the water-borne coating is released to sales, so far as all values lie in the prescribed range.

Some customers of the water-borne coatings manufacturer will request a certificate. In case a test report or manufacturer's test certificate is requested, the values of the external specification are entered. Which values are considered to be external and which are internal specifications must be established by the testing position in cooperation with the producing enterprise and the one entrusted with technical application.

13 Using Water-Borne Coatings

13.1 Substrate Pretreatment

In this chapter the significant processes that are necessary to prepare the substrate for the coatings process will be discussed. The expression "preparation" was selected since very generally the removal of interfering substances (rust, grease, dirt, salts, old coatings materials, etc.) are included, as is also the application of mostly very thin layers, as for example with phosphating and chromating. The goal of substrate preparation is always to attain as intensive a bond as possible between substrate and coating, and thus strong adhesion, since this is a significant condition for the optimal performance of the coating.

Table 17 Pretreatment Methods and Degree of Effectiveness [172]

Effectiveness / Process	Degreasing Effect	Solids Removal	Storage Protection	Suitability as Paint Substrate
Solvent	++	–	–	+/–
Neutral Cleaner	+/–	+/–	+/–	+/–
Alkaline Cleaning with Passivation	+	+	+	+
Alkali Phosphating with DI Water	+	+	+/–	+
Rinse	+	+	+	++

To be noted in general in selecting suitable pretreatment process, is how demanding or "critical" the substrate is and how sensitive the coatings system is to imperfect substrates, what demands are made on the coating and what life expectancy is expected. To begin with it can be established that aqueous products are usually more sensitive to unfavorable substrates than classical solvent-containing paints. The higher the performance capability should be, the more intensive the pretreatment needed. Since very special methods are often required, only an overview can be given at this point. Of key importance is the substrate metal, but the other frequently coated substrates of wood, synthetics and mineral substrates are considered.

13.1.1 Metallic Substrates

13.1.1.1 Cleaning and Degreasing

In the trade realm, various mechanical processes have prevailed, which range from simple brushing of loose dust to removing rust with a wire brush (manual or power) or by grinding (grinding equipment, pneumatic wire brush) to the various blast processes [173]. With the latter processes, liquids (e.g. high pressure water, possibly with the addition of auxiliaries) or solids (metal, silicon carbide, corundum, glass, synthetics) [174, 175] can be used as the blast media, depending on the task at hand. Quartz sand is in general not used much anymore due to the danger of silicosis. A further method is flame treatment, in which the substances to be removed are first burned and than removed mechanically. Next to the required degree of purity, the adhesion of the paint and its thickness as well as its mechanical properties are decisive for the selection of the method. Thus for example, dispersion paints are occasionally criticized, that because of their thermoplasticity (softening on heating), they are difficult to remove. Simple mechanical processes usually do not suffice as primer pretreatment when aqueous, especially dispersion-based systems are worked with. Residual traces of rust act as nuclei for subsurface rusting and thus weaken the corrosion protection effect from the beginning. The recommended degree of rust removal Sa 2 1/2 (bright metal) can only be attained by blasting, which is a costly method both from the standpoint of apparatus and with respect to occupational safety and environmental protection. The disposal of the used blast media frequently creates problems today, when the old coating contains chlorinated binders and heavy metals, as is not infrequently the case. Mechanical processes are however especially effective for the removal of rust and paint residues for renovation painting, however they strongly attack the metal surface. For initial coatings, especially where optimal optical properties of the coating are concerned, chemical methods are better suited [173, 176]. Degreasing is of great significance with the use of water-borne coatings. It must fundamentally precede the application of such products. To be sure, aqueous systems display differing behavior, which is connected with the binder structure and the solvent content; wetting is however fundamentally difficult. There are several possibilities with respect to the process for degreasing. A trend, which is attributable to health and environmental effects, is very clear: The use of halogenated hydrocarbons is constantly decreasing in significance despite high cleaning capability and easy handling (not combustible). To be sure, there are areas where these products can not be dispensed with for technical reasons. The processes used are above all bath cleaning, but also spray and vapor degreasing, which must be carried out in closed facilities [177]. Vapor degreasing is especially effective, since here pure solvent, which condenses from the vapor phase on to the cold metal surface, is used for cleaning. Surfactants (anionic, nonionic) are used above all with aqueous cleaners as well as alkaline or acidic phosphate cleaners, sometimes also containing carbonates or borates. Various cleaners are offered for diverse grease and oil types, so that for many problems special solutions can be found [178, 179]. The processes consist of increasing the efficiency of numerous operations, where again principally baths (supported by the use of ultra sound blasting) and spray applications, sometimes also carried out at elevated temperatures are used. Corrosion protection must always be noted when using aqueous cleaners, which frequently mandates the use of corrosion inhibitors. A significant point is the subsequent complete separation of oil and grease from the cleaning mixture, so that the cleaning effectiveness is maintained for an extended period. In addition to oil separators and centrifuges, ultrafiltration represents a very effective but costly method, which will increase in significance.

In addition to oil and grease, rust and cinder can also be removed chemically [173]. While aqueous cleaners in combination with complexing agents can suffice for light rust, pickling baths of acids (hydrochloric, sulfuric and phosphoric) or lyes must be used for firmly adhering layers. The procedures operating at elevated temperatures are to be sure very effective due to the aggressiveness of the media, subsequently however, a very intensive rinsing of the treated surfaces must be carried out, to avoid re-rusting. The use of phosphoric acid causes the least problems (with respect to damaging the environment by acid vapors, pollution, re-rusting) and by forming thin phosphate layers can contribute to corrosion protection. In the cleaning of zinc and aluminum, one can proceed in principle as with steel. Among aqueous cleaners however only the alkaline products (e.g. phosphates in combination with surfactants) are suited. With aluminum the ever present oxide skin must frequently be removed. This can occur with alkaline cleaners and pickling baths, which as a rule are established with caustic soda. Acid baths (HF/HNO_3 or HF/H_3PO_4) or combinations of alkaline and acidic baths ($NaOH/HNO_3$) are also used for zinc alloys. Alkaline cleaners are also suited for zinc.

13.1.1.2 Phosphating, Chromating

The processes mentioned thus far, consist of the removal of substances, that can disturb the coating. However the application of materials that improve corrosion protection and adhesion can also belong to substrate pretreatment [180]. This concerns in large part industrially used processes. The most significant is phosphating, whereby zinc phosphating is meant above all. The most important chemical reactions for the treatment of steel with an acid phosphate solution are [173]:

Pickling Reaction

$$Fe + 2\,H^+ \rightarrow Fe^{2+} + H_2$$

Hopeite Formation

$$3\,Zn^{2+} + 2\,H_2PO_4^- + 4\,H_2O \rightarrow 4\,H^+ + Zn_3(PO_4)_2 \cdot 4\,H_2O$$

Phosphophyllite Formation

$$2\,Zn^{2+} + Fe^{2+} + 2H_2PO_4^- + 4\,H_2O \rightarrow 4\,H^+ + Zn_2Fe(PO_4)_2 \cdot 4\,H_2O$$

The deciding step is the precipitation of the finest possible crystalline phosphate on the metal surface and the formation of a thin, tight protective layer. Especially in the automotive industry, this pretreatment in combination with CED primers has brought a clear improvement of the corrosion-protection properties [181, 141]. Further process optimization can be attained by reduction of the zinc content (low zinc process, favored phosphophyllite formation) and by the addition of nickel, manganese or calcium. Phosphating consists of a multi-step process, which in addition to cleaning, rinsing before and after (DI water and passivating after-rinse with chromium III/VI solution), includes also a special activation of the metal surface (addition of titanium phosphate solutions to increase the formation of crystallization nuclei) [182]. The real phosphating occurs in the dip or spray process. A further method is iron phosphating which is carried out with less acidic solutions of phosphoric acid and alkali phosphates. The iron phosphate that forms after release of iron ions precipitates on the steel surface and forms a protective layer, which however does not offer the good corrosion protection of zinc phosphating.

Zinc phosphating can also be carried out on zinc and aluminum. In both cases the above mentioned hopeite is formed [173]. Phosphating of aluminum is somewhat costly, since it must be supported by the addition of accelerators (nitrites, chlorates, peroxides) and fluoride ions (dissolving the oxide layer, complexing of aluminum to avoid AlPO4 precipitation). Kryolith is formed by the addition of fluoride.

Chromating is a further type of surface treatment [173], which is especially used with aluminum and can be carried out by immersion or spraying. Hydrofluoric acid, phosphoric acid and chromic acid are used as the important components. Hydrofluoric acid again serves to dissolve the oxide layer, while as with aluminum and phosphate,the chromium [III] formed from chromic acid, forms the insoluble products aluminum and chromium phosphate which develop a protective layer on the aluminum surface. One distinguishes between green and yellow chromating, whereby the green color results from the chromium phosphate and the yellow color is derived for certain quantities of chromate. Similar to phosphating, chromating includes several steps: degreasing, rinsing, pickling, rinsing, chromating, rinsing and subsequently rinsing with DI water.

A further pretreatment method which is occasionally recommended is the so-called rust conversion [173]. With this the rust is not removed , but chemically converted. Phosphoric acid or organic acids are recommended as reaction partners for the rust. The desired effect can only be achieved with precise observance of the stoichiometric ratios, which however in practice can not be realized [183]. As a result this process is not considered suitable. The use of tannin to complex the iron from rust usually does not produce the desired effect. Especially in combination with water-borne coatings, no successful corrosion protection can be achieved with incomplete removal of the rust present or with too high a dosage of the reaction partners.

13.1.2 Non-Metallic Substrates

Based on the total surface of substrates to be coated, mineral substrates have a significant share. While decorative effects are frequently the point of focus, the protective function is significant with for example concrete, which is painted to an increasing extent. Here too, it is necessary to observe diverse criteria, before a coating can be applied [184,185]. In contrast to metallic surfaces, with most mineral substrates, the surface integrity must always be checked. Loose and friable spots must be removed and improved, since they can not be kept together even with a new coating. In the case of cracks it must be determined what type is present. Cosmetic cracks should be primed and spackeled, while constructive cracks must be filled with an elastic joint compound before painting.

The integrity and adhesion of the old paint must also be checked. Rinsing and priming may be necessary. With good dispersion paints, a precoat with the thinned paint usually suffices. To be removed in addition are impurities and deposits such as dirt, grease, oils (e.g. release agents on concrete), blooms such as fungi, lichen and algae. Cleaning can be carried out with brushes and also with high-pressure cleaning, flame and sand blasting (concrete) or scraping. An important aspect is the absorbency of the substrate, which can be controlled and reduced by the use of penetrating primers. In the case of concrete, damage that goes deeper and has attacked the reinforcement must be restored. Especially in construction, strict regulations must be followed; the work can only be carried out by licensed professionals with approved restoration systems. Painting, which above all must protect against the effect of acidic agents and stem the diffusion of CO_2, is the final activity.

The pretreatment of plastics [126] again includes the removal of interfering substances. To this belong above all products, that were used as auxiliaries for the manufacture and processing of the plastics, but also grease and dirt in general. The use of aqueous surfactants or alkaline cleaners, often at slightly elevated temperatures, is favorable, while the use of solvents can be critical, since stress cracks can result with many plastics. A problem with the cleaning of polymeric materials is the strong tendency for building up electrostatic charge as a result of the low conductivity, which especially favors the deposition of dust. It can therefore be sensible, to blow off already cleaned surfaces again with ionized air before the painting process. An alternative is finishing or subsequent coating with anti-static agents. Adhesion between paint and substrate is a further problem. Plastics present unfavorable conditions for the development of good adhesion as a result of their chemical nature (especially with non-polar polyolefins) and the usually flat surface structure. A solution here can be roughening by scraping or blasting or working with adhesion promoters. Very effective are surface treatments, that effect an increase in polarity and thus the number of contact points. Such procedures are e.g. flaming, treating with chlorine, immersion in strongly-oxidizing acids (e.g. chromosulfuric acid) and especially the use of corona discharge or plasma.

A significant disadvantage, that exists for the application of aqueous systems (paints, glazes, primers) on wood substrates in comparison with solvent-containing systems, is the pronounced roughening of the substrate as a result of the swelling of the wood fibers. Thus, careful sanding with subsequent complete removal of the dust is a fundamental precondition for satisfactory painting. Resin constituents, that issue from the wood, must be removed and (natural) defective areas must be compensated. Further, just as with other substrates, impurities and dirt must be eliminated.

13.2 Applications Processes

13.2.1 Brushing Application

Wherever fixed, small area, strongly structured and complicated objects are to be painted, brushing is the oldest and simplest procedure, the method of choice [13]. Thus, the glazing varnish for a base board as well as the corrosion protection painting of a tubing bridge are applied with brush and mop. In spite of the relatively high labor expenditure and limited precision of the process (layer thickness, surface structure) the manual approach is favored, since alternative methods such as immersion and spraying are eliminated for analogous reasons.

Application by brushing has increased in significance with water-reducible systems, since water-borne coatings are used as alternatives to classical solvent-containing systems in increasing magnitude. To be sure, fulfilling the requirements of this supposedly simple process for aqueous systems is not so simple [186]. The difficulties lie above all in paint flow and drying where the binder type plays a large role [187, 188, 189]. Water-soluble resins in general perform better in this regard than polymer dispersions, which on the other hand offer the advantage of low solvent/film forming agent content. The unfavorable behavior of dispersion paints results from the fact that with relatively high rest viscosity (flow limit) they quickly lose viscosity with increasing shear force. As a result the familiar resistance of "tougher" paints is lacking with brushing, which easily leads to pronounced streaking, so that the necessary film strength cannot be attained. Because of the high rest viscosity, the paint material does not flow fast enough after brushing, much more it almost sets up, so that brush marks remain visible. Thus the paint surface does not offer the familiar appearance of classical systems with respect to smoothness and gloss. This undesired behavior of dispersion binders can be countered by the use of specially introduced products, frequently in combination with appropriate rheological additives (especially polyurethane thickeners, see chapter 8.5).

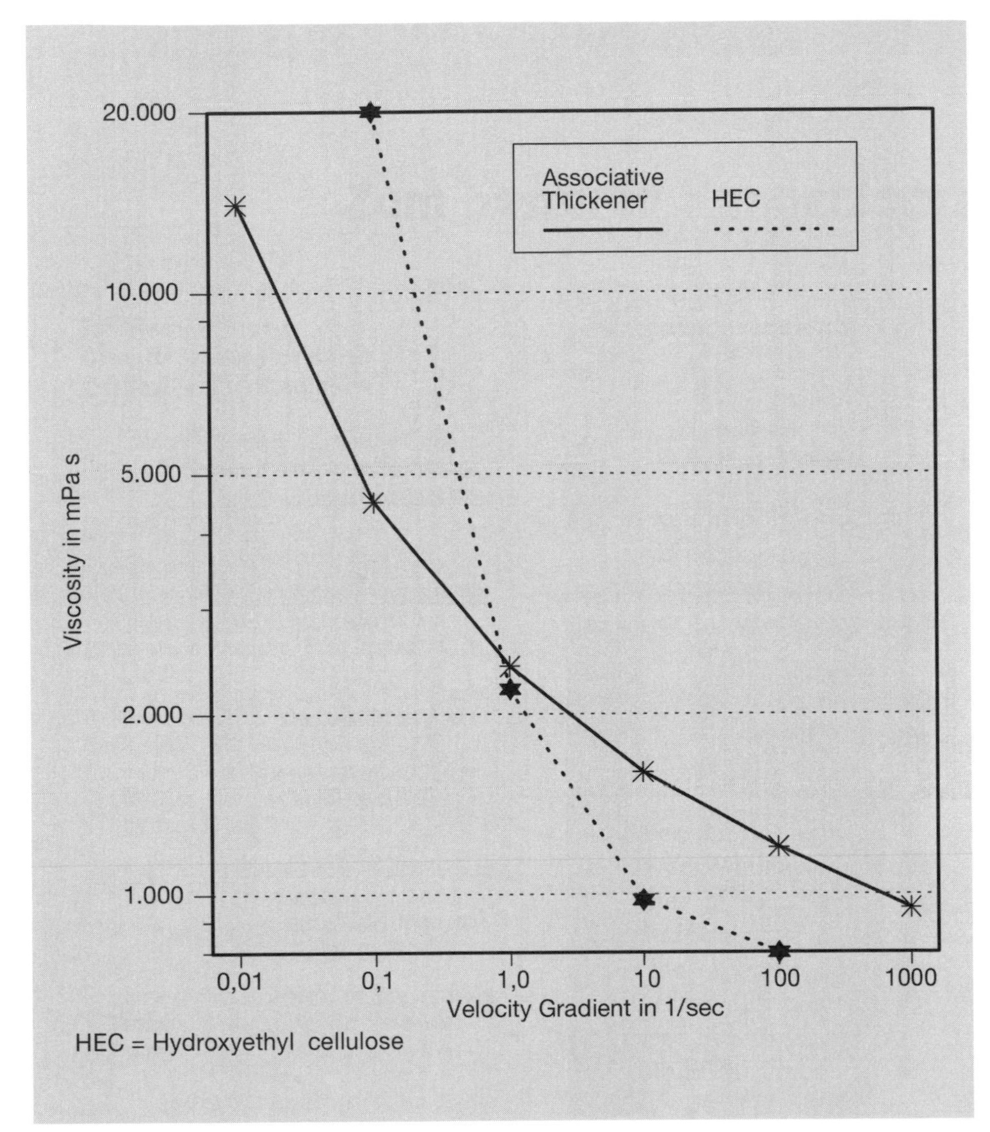

Figure 18 Rheology Curves of an Emulsion Paint as a Function of Thickeners

By such measures the rheology curve of a dispersion paint is flattened and a more Newtonian behavior is attained: The lower rest viscosity enables better flow, and the lower viscosity decrease with shear force creates a certain toughness, which reduces the danger of too strong streaking. A further difficulty is the relatively fast drying. Dispersion paints often display a too short open time, since drying proceeds from top to bottom. In addition there is the irreversibility of the film formation with dispersions. This creates the problem for the painter, that on repainting already partially dried parts, the film is torn up and the smooth surface is disturbed. Smoothing out of the paint is thus made more difficult. These effects can be ameliorated by the addition of low-volatile auxiliary solvents, that slow the release of water. The water-borne coatings offered today have significant advantages compared to before, however performance equivalent to the classical coatings has not yet been attained. Based on the formulation, a mixture of dispersion and water-soluble resin can offer advantages, if the higher solvent content is allowed for. Of importance also is the use of suitable brushes. In addition to expensive long-haired China bristles, specially formed synthetic brushes with less tendency to gum up have prevailed. Because of the danger of too much streaking, water-borne coatings should not be brushed too intensively, rather the paint should be more laid down. For larger surfaces, as e.g. doors, special procedures are recommended that reduce the problem of tearing up the paint surface [190]. If the optical impression does not play such an important role, as e.g. with corrosion-protection paints, the technology used with chlorinated rubber can be used.

13.2.2 Roller Applications

Rolling is for many years the applications procedure for classical dispersion paints, used by both the trade and do-it-yourselfer. The technology is relatively simple and thus easily learned by the layman. In addition it is more efficient (with relatively large areas such as walls and ceilings) and more uniform than brushing. A disadvantage of rolling is the relatively weak wetting of the substrate, since the material is laid down rather than worked in. The substrates should preferably be flat and uncomplicated, or brushing is preferable. Usually a definitely structured surface is obtained with a roller, where the effect is strongly dependent of the rheological behavior of the paint or coating as well as the type of roller. With the latter, the palette ranges from long-napped to plush to largely flat rolls of foam. Materials are above all wool, perlon or nylon. A applications disadvantage of rolling is the tendency to splatter, which implies reduced efficiency (material loss) and to messiness. Splattering is reduced by the use of suitable additives. In the last few years a very low-splatter "firm" paint has been offered. The material has a pudding-like consistency and is only made sufficiently low in viscosity for distribution on the substrate by the shear force of rolling. After the shear force is withdrawn, a high viscosity spontaneously reforms, so that splattering is eliminated. The tendency to splatter is also dependent on the type of roller selected.

13.2.3 Spray Application

In this chapter, the conventional spray processes, that are extensively used in the trade area, will be discussed [191]. To be sure the technology has been developed so far that is practiced on an industrial scale without human participation. For example in the automotive industry entire paint lines are equipped with paint robots and high-speed rotation atomizers. The electrostatic processes used there, above all will be described in chapters 13.2.4. and 16.1. [192, 193].

Compared with the typical techniques of trade application, spraying offers the special advantage of significantly faster and smoother processing, which makes it the method of choice specially with large simply-structured objects [13]. It naturally requires a larger equipment expenditure. In general, water-reducible systems are well suited for spray application [194]. In contrast to brushing, there are hardly rheological problems. The coatings producer can establish correct behavior by the use of suitable thickeners. Problems can occasionally arise on thinning to increase the spray capacity: A significant change in rheology comes sooner, and therefore unexpectedly for the processor, than with solvent systems, and insufficient rigidity can result. In contrast with the broadly accepted opinion, that aqueous systems dry especially slowly (see chapter 14), one is frequently confronted with the opposite on spraying. Thus a buildup of a crust of dried or coagulated paint material can occur in the area of the nozzle finally leading to blockage. Such problems must be considered during formulation development. This applies also to the drying of the sprayed paint. With unfavorable atmospheric conditions, drying, as a result of the fine dispersion and large surface area, can advance sufficiently on the way to the object to be painted, so that the paint is already partially dry on hitting the surface and the formation of an integral film is made difficult [172, 187].

There are several points to be noted when spraying water-borne coatings, which were already discussed with the manufacture of coatings (see chapter 11). The shear force to which aqueous systems reactive sensitively must be mentioned. Especially in airless equipment, which affects the paint strongly at high pressure, coagulation can result. The coatings manufacturer must already be cognizant of these effects when selecting the binder. To be sure, the spray equipment must be laid out correspondingly. The materials must be inert and plated metals and rusty iron or steel must generally be avoided. Stainless steel, polypropylene or teflon are suitable materials for lines, containers, pumps, sprayers and seals. Catastrophic consequences can result from changing aqueous and solvent-containing products in spray equipment if it is not carefully cleaned. Contact with product residues can lead to coagulation of the water-reducible binder and in the worst case, to severe damage of the spray gun or pump (airless process). Also with exclusive use of aqueous products, thorough cleaning with water is necessary after application, since subsequent cleaning even with the use of solvents is hardly possible because of the polymer properties of dispersions. With water-soluble resins the behavior is more favorable, since even after drying, a certain solubility remains. In conclusion it must be made clear that spraying water-reducible paints brings nothing fundamentally new for the processor, but a certain getting used to and discipline, as well as maintaining different playing rules is needed [195].

In general, spray applications always involve the atomization of the paint material which is conveyed to the substrate where the droplets coalesce to a film [141]. The various techniques, that have been developed over time, serve to improve certain aspects of spraying: Increasing the coating capacity, smoother coating, the possibility to process higher-viscosity products, minimizing spray fogs and attaining a higher material use efficiency [141, 173]. Some processes will be discussed in this chapter; others in chapter 13.2.4.

The conventional spray process is compressed air spraying, where low and high pressure spraying can be differentiated. The process consists in compressed air accelerating the paint material and causing its atomization. The degree of atomization of the paint depends on the weight ratio of the stream of paint to the stream of air, so that in increase in the air stream leads to a reduction in the median particle diameter. A reduction in the paint viscosity has the same effect. The low-pressure process is inferior to the high-pressure process in that the air pressure (0.2 – 1 bar) and stream weight are less and the paint particles are coarser. Indeed a blower suffices as power which can be especially advantageous in construction. For high-pressure spraying , pressures up to 6 bar are required. In general water-borne coatings require higher pressure than classical coatings to be atomized in comparably small droplets. Further differences in spraying technique are in the different ways of paint delivery. Cups on the spray gun are suited for small quantities and quick material change. The reservoir, the cup, which can be a siphon, pressure or flow cup, is located directly on the gun For larger quantities, the material is located in a separate container and is transferred from this to the gun. The development of spray fogs represents a disadvantage insofar as a certain amount of material remains unused, and subsequently condenses as paint sludge and must be disposed of or better, returned after work up (see chapter 17.4). In addition, there is the danger, that spray fogs will fall on already painted surfaces and lead to film and surface disturbances. In sum, conventional spraying is a very capable, flexible method with limited equipment requirements, which to be sure requires some experience on the part of the operator and has some disadvantages with respect to paint utilization (overspray). The subject of disposing of paint sludge and recycling will be reported in chapter 17.4.

A further development is airless spraying, where the paint material is directly, e.g. without use of air as transport medium, accelerated and atomized. The appearance of interfering fogs is strongly reduced here. As a result of the reduced atomization however, this advantage is however tied to a sharply contoured spray beam and the development of coarse paint droplets, so the paint surface does not present the high quality of conventional spraying. Finer atomizing can be achieved by increasing the throughput and widening the angle of the opening in the nozzle. A further advantage lies in the high efficiency (good material utilization, high spray capacity, high film thickness in one coat) and the ability to apply also higher viscosity coatings materials. Of further advantage is the good wetting of pores and depressions in the substrate surface. The expenditure for equipment is relatively large, and there are higher demands on occupational safety due to the high pressure [13]. Of advantage is being able to work directly out of the supply container. The airless process is favored above all for the coating of larger surfaces, where extreme demands are not made for the optical appearance of the paint. Aqueous systems in principle can be processed without problems. To be noted when formulating, is that the rheological properties persist also with the high shear forces that occur here. Good rigidity on vertical surfaces is necessary. Further the paint material must be resistant to the mechanical effects which occur on pumping. On the other hand the exclusive use of inert materials of construction must be observed.

The so-called airmix process represents a combination of the airless technique and air pressure spraying. With this the paint is again directly, i.e. hydraulically conveyed and atomized. After leaving the nozzle, an additional atomization occurs by the introduction of pressurized air. In this way more viscous coatings materials can be sprayed, while maintaining a relatively fine dispersion and a less sharply contoured spray beam. Drying of the paint material is accelerated as a result of the air introduction. In summary the method combines to a large extent the advantages of the airless and pressurized air sprays. The equipment cost is however significantly higher, since both an airless pump and an air compressor are needed. Thus this process should be considered for manufacturing, the equipment costs do not permit operation in construction.

13.2.4 Electrostatic Spray Processes

Fundamentally, two techniques can be distinguished with electrostatic spray processes [173, 196]. With the one, atomization of the paint is achieved with purely electrostatic methods (spray slit, bell, disc). Since water-borne coatings can be processed in this way only with difficulty, it will not be discussed here. In the other, the actual atomization occurs mechanically with the techniques presented in chapter 13.2.3. or with the rotation equipment common exclusively in series coating such as high rotation bells and discs. Subsequently an electrostatic field is established between the atomizer and the work piece, whereby a target-oriented movement of the charged paint droplets on the object is obtained. With this modification one also achieves a significantly higher degree of efficiency with conventional methods. The process is especially advantageous for parts of complicated structure with areas difficult to approach. The so-called wrap-around effect leads to even the back of the part being partially coated.

Charging the paint, which is difficult with water-borne coatings due to the high conductivity can be accomplished in two ways: the

- direct charging by conduction (electron or ion conduction) or the
- charging in the corona field, also called external conduction.

With electrostatic airless or pressurized air spraying, only the second method, – i.e. charging outside of the gun – can be used. If the charging were carried out inside the spray gun, the high tension would be destroyed by short circuiting. Isolating the material delivery, which could prevent a short circuit, is not possible for safety reasons (too great a danger for the operator) [197]. For these reasons, an external charging is carried out with the help of so-called horns, that are fitted to the front of the spray gun. After atomization, the paint particles pass through the electric field built up by the horns becoming charged by taking up air ions and are then steered by the lines of force the object to be coated. The effect, above all the wrap-around effect mentioned, is not as strong with water-borne coatings as with conventional systems for two reasons: External charging is not as intensive as the direct method, and the atomization of water-borne coatings usually occurs at higher pressures, so that the steering of the particles in the spray direction can be greater than the field direction. Thus it is advantageous to orient the electric field and the spray beam in the same direction. The processed described here can be used in both manufacturing by manual procedures and in line coating (e.g. in the automotive sector), when parts that are difficult to approach are concerned. Here the use of spray robots, which do not differ fundamentally with respect to the spraying techniques of manual operation, but bring higher productivity and especially better reproducability of the coatings process, is becoming ever more fashionable.

High rotation bells and discs are tools that are especially suited for the industrial painting of flat pieces [141]. The paint material is supplied to the interior of the bell or discs and is forced outward by the centrifugal force created by the extremely fast rotation (advantageous for water-borne coatings ca. 30,000 RPM). In this way an atomization is attained that is not only finer but in addition displays a narrow particle size distribution. In sum a high paint throughput is achieved. Consideration must also be given to the high electrical conductivity of the water-borne coating. If one operates with direct charging by conduction as with solvent-containing coatings, a relatively expensive inflexible cascade system must be used. Here the paint supply and applications unit are strictly separated. Filling the paint reservoir proceeds by way of in intermediate vessel. This is filled while grounded from the supply line and subsequently after applying the high tension, emptied into the actual reservoir. The external charging method is especially easy to operate in the case of frequent color or paint changes [196]. Here the electrodes are usually in the form of rings and attached to the spray head of the high-rotation atomizer. In spite of the aforementioned low effectiveness of this charging method, a target-directed air stream is achieved by the support of the subsequently introduced forced air. The achieved degree of effectiveness of over 90% comes close to that of classical coatings. Of great significance for successful work is the precise coordination between applier, paint and facilities manufacturer. Precise conceptions of the total system must exist when planning so that arrangement, atomizer rotation speed, throughput, spray distance, air and other parameters can be established exactly [193]. The accelerated drying of the paint material, with the already mentioned disadvantageous effect on film building and surface structure, is especially difficult with dispersion-containing products because of their very fine atomization. Further deposits on the rotary bell can occur. These difficulties can be countered by the addition of water retention agents or mixing with water-soluble resins or by increasing the humidity of the cabinet air [172, 187, 198, 199, 200].

13.2.5 Dip Application

Dip painting is a process that is usually used in industry [13]. Installing and maintaining a dip tank is only justified when a relatively large number of parts is to be coated. While a large quantity of paint must be available for the tank, only a fraction of this quantity is needed for coating an object. The method is advantageous for complicated, angled pieces which cannot be completely painted by the spray process. Technical requirements are not as high as with spraying, however precise establishment of the transport of the object to be painted through a dip tank, whereby above all the entrance and exit phases are decisive, is required. A formulation that is correctly standardized with respect to rheology is necessary to achieve an irreproachable coating without runs or curtains. Often a binder combination of dispersion and water-soluble resin is suited for aqueous systems. Limitations are imposed on classical dip application by the limited increase in film thickness and the problem of edge covering. Fundamental improvements in this respect are achieved by the use of modified dip processes such as electodeposition painting and autodeposition. In addition to facile applicability, the paint must display very long storage stability, since the average residence in the dip tank is often very long as a result of the aforementioned quantity ratios. Thus neither a gradual phase separation nor skin formation or gelation can be allowed to occur. Frequent controls (viscosity, solids content, pH, surface tension, etc.) of the dip paint ensure irreproachable functioning. Corrections can always be made on refilling the tank. To be noted is that auxiliary solvents sometimes vaporize faster than water and thus can escape faster. In addition to these measures, which concern more the formulation and its supervision, the dip tank itself must be suitably equipped. Constant agitation of the paint, avoiding static blind angles and careful enclosure of the tank on extended shutdown are required to maintain homogeneity and to prevent skin formation. The problem of irreversible film formation, especially with polymer dispersions, has already been referred to numerous times. The parts to be painted must be well cleaned before dipping, to prevent the constant introduction of impurities and thus altering the dip paint. Usually the dip tank is part of a production line, in which a cleaning arrangement before and a drying train after are included. In practice, it is possible to consider the operation of a dip tank with water-borne coatings with careful maintenance of the aforementioned measures throughout as problem free. The high fire risk in a tank with solvent systems is reason alone for the increased use of water-borne coatings in dip applications [173]. A further aspect is naturally the reduction of solvent emissions. Precisely with respect to environmental problems, dip painting in general offers the advantage of achieving a high degree of effectiveness, whereby above all low paint wastes result [201].

13.2.6 Electrodeposition Application

While water-borne coatings are relatively new systems, the electodeposition coatings are considered more as classical [76, 202, 203, 204, 205]. They have already been used for 30 years in industrial painting and show a constant growth rate in quantity [206]. Similar to the transition from conventional to electrostatic sprays, this special type of dip coating represents a considerable improvement in the conventional method. To be named as advantages are:

- No problems with runs and curtains
- Irreproachable edge coverage by achieving a uniform film thickness on the entire object to be coated
- Problem-free coating also in hollow places (wrap-around)

Since principally only water-soluble binders can be used with these special processes, this is considered environmentally friendly. To be sure a higher facilities and materials cost, and thus higher investment costs and quite demanding supervision are required for extensive automation and continuous operation. In the end there is a coating with excellent corrosion protection and good surface properties.

The important process step of electodeposition coating is the migration of the electrically charged binder molecules to the likewise charged object to be coated and their precipitation on the surface by coagulation [207]. Depending on the charge one distinguishes between anodic electrodeposition (AED] and cathodic electrodeposition (CED) coating. The AED method was developed first, while in recent years the CED process has penetrated the automotive sector especially, due to its advantages and is generally preferred today [208, 209, 210]. Both techniques however have advantages and disadvantages and thus possess their own justification. In the following, the mechanisms will be briefly described and subsequently the detailed progress of the total process will be discussed.

- *Anodic Dip Coating:* Here the object to be coated forms the anode and the tank, or better immersed electrodes, form the cathode. The molecules of the water-soluble binder, which are present as salts, in this case as anions, migrate to the anode:

Salt formation: $RCOOH + NH_3 \rightarrow RCOO^- + NH_4^+$

Anode reaction: $2\ H_2O \rightarrow 4\ H^+ + O_2 + 4\ e^-$

$RCOO^- + H^+ \rightarrow RCOOH$ (water insoluble)

The binder coagulates in the area of the anode, since it reacts with protons from the aqueous electrolysis and thereby again becomes water insoluble and precipitates on to the metal surface.

- *Cathodic Dip Coating:* Since in this case the article is the cathode (the anode is formed from graphite or stainless steel), a binder type is needed which becomes water soluble by the formation of cations. This effect is achieved by binder molecules with amino groups which react to form soluble ammonium ions on addition of carboxylic acids (formic, acetic).

Salt formation: $R_2NH + RCOOH \rightarrow R_2NH_2^+ + RCOO^-$

Cathode reaction: $2\ H_2O + 2\ e^- \rightarrow 2\ OH^- + H_2$

$R_2NH_2^+ + OH^- \rightarrow R_2NH + H_2O$

The reaction with hydroxyl ions in the region of the cathode reforms the insoluble binder, so that coagulation and precipitation occur.

Independent of the sign of the charge, there are further details of the precipitation mechanism that are important for the success of electrodeposition coating. Thus on the one hand the binder ions migrate in the direction of the electrode (electrophoresis), while the aqueous phase flows in the opposite direction(electroosmosis). This results in the coating having an astonishingly low water content already on exiting the dip bath. The aforementioned uniform film thickness, that is achieved on the entire object to be coated, results from the following circumstance: The precipitated paint has a relatively high electrical resistance, so that the rate of precipitation decreases with increasing film thickness. Thus in the course of the dip procedure, the places where originally little material was precipitated are also seized. The wrap around can be optimized by certain fine adjustments that will not be discussed here.

The CED process has achieved its preeminence through its better corrosion protection with lower film thickness and better wrap around. Further the applied voltages of 20–50 volts are significantly lower than with the AED process with 50–300 volts. It is never the less to be noted that a not problem-free chromate rinse is used here, which was not the case originally with the anaphoretic process. The latter offers above all less technical expenditure and offers with respect to the binder used, not only a cost advantage, but also more freedom in formulating. In addition, coatings with a certain flexibility can be achieved, while the CED paints are more brittle.

In closing a few observations about the entire electrodeposition paint procedure, which in addition to the dip procedure consists of further individual steps. Prerequisite for the process is an intensive pretreatment of the metal substrate, which consists of rust and grease removal (alkaline or with solvents) [126, 141]. Usually the use of phosphating is included.

As with conventional dipping, the dip paint must be kept homogeneous by constant agitation (stirring, pumping). Important values such as solids content, amine or acid content, conductivity, pH and temperature must be regularly monitored. Since heat is released with electodeposition coating, a cooling circuit with heat exchangers is necessary to maintain constant temperatures (typical temperature: 20 – 35 °C). The paint must be constantly filtered to remove small particles, as for example metal shavings from automotive chassis, that could damage the surface structure of the coating. Two effects are accomplished by ultrafiltration included in an additional circuit: Since, in contrast with the conventional process, the solid components (binders, pigments) are preferably precipitated in electrodeposition coating, the tank contents are gradually depleted of these substances. This is equalized by the ultrafiltration, where water and other low-molecular weight components are separated, which would otherwise build up in concentration. The ultrafiltrate is not thrown away, but is used as rinse water in the posttreatment. In this way, the degree of efficiency of the paint can be significantly increased. The crude concentrating with help of the ultrafiltration naturally does not suffice to maintain the solids level, more often additions must be made. To prevent the enrichment of acids released on precipitation (pH shifting), partially-neutralized materials can be used for the addition. These are often highly concentrated and require premixing in a separate container. An alternative is including a so-called anolyt circuit (electrodialysis), in which the acid is separated, so that normal standardized fully neutralized paint can be added. In addition to these supplementary circulatory systems, which are essential for proper functioning, a holding tank must be installed, which can hold the total paint material during repairs or cleaning.

A rinse process follows the actual paint application. Here in a preliminary procedure, the unprecipitated paint residue is rinsed with the aforementioned ultrafiltrate. Subsequently a reactive rinse with chromate-containing solution takes place, followed by a final rinse. Following a pre-drying of the coating, the baking process at temperatures of up to 180°C completes the process. In recent years further efforts have been made to attain improvements especially in cathodic electrodeposition painting [211, 212, 213]. These concern the paint itself and the process with regard to quality and environmental protection. Thus the elasticity and stone chip resistance of CED coatings especially needed improvement, but also the corrosion protection. Additional points are the film thickness distribution, the processing safety, baking temperatures, as well as further reduction in the auxiliaries used and the paint wastes (see also chapter 16.1.1.2.).

13.2.7 Autophoretic Coating

This also concerns a dip process which is modified with respect to the precipitation process [214]. Frequently it is mentioned in the same breath as electrophoretic methods, although it rests on a different mechanism, which proceeds differently and leads to different results. An important difference with electrophoresis is that the coating material is chemically (thus also chemiphoresis), i.e. without the use of electric current, precipitated on the metallic substrate, especially steel. Also in contrast with anaphoresis and cataphoresis, acidic anionic polymer dispersions are used as binders. Hydrofluoric acid is especially used to adjust pH, whereby the bath also contains iron fluoride. An oxidation material (hydrogen peroxide) is required as an important additive. In addition, black pigments (in special cases also other color tints) can be included. The precipitation process [215] is initiated by the iron [II] ions which are released from the metal surface by the effect of the acid, which destabilize the diluted latex and cause coagulation. The coagulate precipitates on the metal surface. Subsequently the iron [II] is oxidized to iron [III] which diffuses out of the growing paint film to the environment where it is complexed by fluoride ions. The precipitation is thus closely tied to the evolution of iron [II] ions. The rate of precipitation depends on several factors (latex concentration, agitation speed, salt concentration), whereby film thicknesses of 20–30 μm are achieved in 1–2 minutes. Since the rate of precipitation decreases with increasing film thickness as with electodeposition coating, here too a uniform paint application is obtained. The problems that occur with conventional dipping, such as poor edge coating and running, are avoided. The coating of hollow spaces is even better than with the CED process.

The process steps accompanying the precipitation process are the following [216]. First the metal surface must be cleaned by intensive treatment (alkaline cleaning and degreasing at 60-80 °C; rinsing with fresh and DI water for the complete removal of foreign ions). Since a bare metal substrate is required, phosphating is not possible, but also not necessary. After precipitation, which is carried out at room temperature, the coating is maintained briefly in a very moist atmosphere to reduce cracking susceptablilty (shrinkage as a result of incipient drying). In contrast to electrophoresis, no osmotic process occurs here, which would lead to dewatering of the applied film, rather about 50% or more water is retained. A rinse with fresh water to remove uncoagulated latex residues and a reactive rinse with a chromium [VI] solution follow. The latter leads to a significant improvement in the corrosion protecting properties. At the end are a predrying to evaporate the water followed by a baking process at about 150 °C, which causes the actual film formation of the dispersion binder. Constant control of the bath is critically necessary (solids content, surface tension, pH and redox potential) as well as constant agitation. Here however no excessive shear forces are allowed because of the sensitivity of the binder. The bath temperature must also be precisely maintained, which is however less difficult since the process proceeds at room temperature and is not exothermic. Autodeposition offers some advantages compared with electrophoresis [217, 218]. To this belong the less intensive pretreatment and the lower energy costs (no electricity, simpler temperature maintenance, lower drying temperatures). The coating thus achieves a corrosion protection without phosphating that is comparable to cataphoretic coatings with phosphating. By using polymer dispersions, relatively flexible coatings can also be obtained. Disadvantages are the limitations in color choice (no light colors are possible) and the broad fixation on iron and steel substrates. Further bath maintenance is difficult and requires great care. Unlike the CED process, the post-treatment cannot do without chromium [VI] ions which is unfavorable from environmental grounds.

In spite of the advantages described, autodeposition has found no great distribution in recent years, its use is more limited to special applications (complicated formed small parts) [215].

13.2.8 Paint Application by Pouring and Rolling

Of the additional applications processes for coatings, flow coating and roller coating will be discussed briefly. In flow coating, the object to be coated passes on a line through a continuous pour curtain. This results from paint material flowing over so called weirs, which are attached to the entire width above the transport line. If no object is under the pouring curtain, the paint material is recaptured underneath and returned to the paint reservoirs. Very small paint losses result, so that the degree of effectiveness is comparably high as with the dip processes. As with conventional dipping, the attainable film thicknesses are limited because of the flowable consistency of the paint, whereby higher application quantities are made possible by a horizonal orientation of the object. To be sure the parts to be painted must be largely, if not completely flat, since otherwise no uniform coating can be achieved. Regulating the coatings thickness is possible by varying the throughput rate and the justification of the pouring weirs. A further limitation of the method is the fact that the paint is more laid on, so that anchoring to the substrate is relatively weak.

Roller coating can also be used with completely even articles, which also gives a very high degree of effectiveness and a very uniform coating. The process is widely used in wood coatings for example. In contrast with pouring, paints can vary greatly, above all also highly viscous paints can be processed. Two methods are differentiated, where an object and application roller move in the same direction (concurrent process) and where movement is countercurrent (reverse process). In the latter the paint thicknesses are higher. In general, front and back sides can be painted in this way. Shear forces and drying conditions must be monitored especially when using dispersion systems.

A further process is used especially in coating and painting paper and cardboard. In this the material is poured on to the very rapidly moving substrate and is subsequently pressed uniformly on the paper with a "knife". In this case very high shear forces occur at the "knife" so that only specially developed dispersions can be used for this application.

14 Drying of Water-Borne Coatings

The drying of water-borne coatings and how to influence it has been investigated by many research groups [219, 220, 221]. The following factors especially influence the progress of drying of water-borne coatings:

- type and amount of cosolvents
- the relative humidity
- the air velocity
- the temperature.

As an example, figure 19 shows the dependance of evaporation of a solvent-water mixture on the relative humidity. Figure 20 shows the temperature-dependent evaporation curves of the water-solvent mixture of water-borne coatings based on a polyester or a polyacrylate. For purposes of orientation, the evaporation curve of butyl glycol is also shown. The representation shows the differing binder-dependent water-solvent retention which is especially pronounced at higher temperatures and results form the different drying rates on the paint surface of polyester and polyacrylate coatings. Film formation of the binder proceeds together with drying, that means together with the evaporation of water and solvents if present, or immediately after [173]. This process can be of a purely physical nature, as for example with polymer dispersions, or can consist of a molecular enlargement of the sort where binder molecules are cross-linked by oxidation, polymerization, polyaddition or polycondensation.

As a rule, the drying of aqueous systems (especially emulsion systems) is more difficult than solvent-containing systems. This is especially true for exterior applications, which are strongly dependent on environmental conditions such as humidity, temperature and wind speed. Thus drying at excessive wind speeds can be much too fast and lead to crack formation, while the same paint hardly drys at high humidity and low temperature. Uniform drying, and thus complete optimal film formation, can be furthered by formulation adoptions as for example the addition of suitable auxiliary solvents [200]. Very important is that sufficient film forming agents be present with progressing water evaporation to guarantee film formation. Constructive measures, such as shielding the object with tarpaulins (often necessary to protect the environment from paint splatter where corrosion-protecting paints are concerned) and protection against the elements simplify the work.

With industrial painting, the problem is very different. The demands on the surface condition of the paint are usually very high, so that optimal application and drying conditions must be maintained. It must be noted that drying, e.g. with spraying, begins immediately on exiting the spray gun, i.e. during the application. It must be assured that at a given point in the drying process, the water content and the ratio of water to auxiliary solvent are always the same.

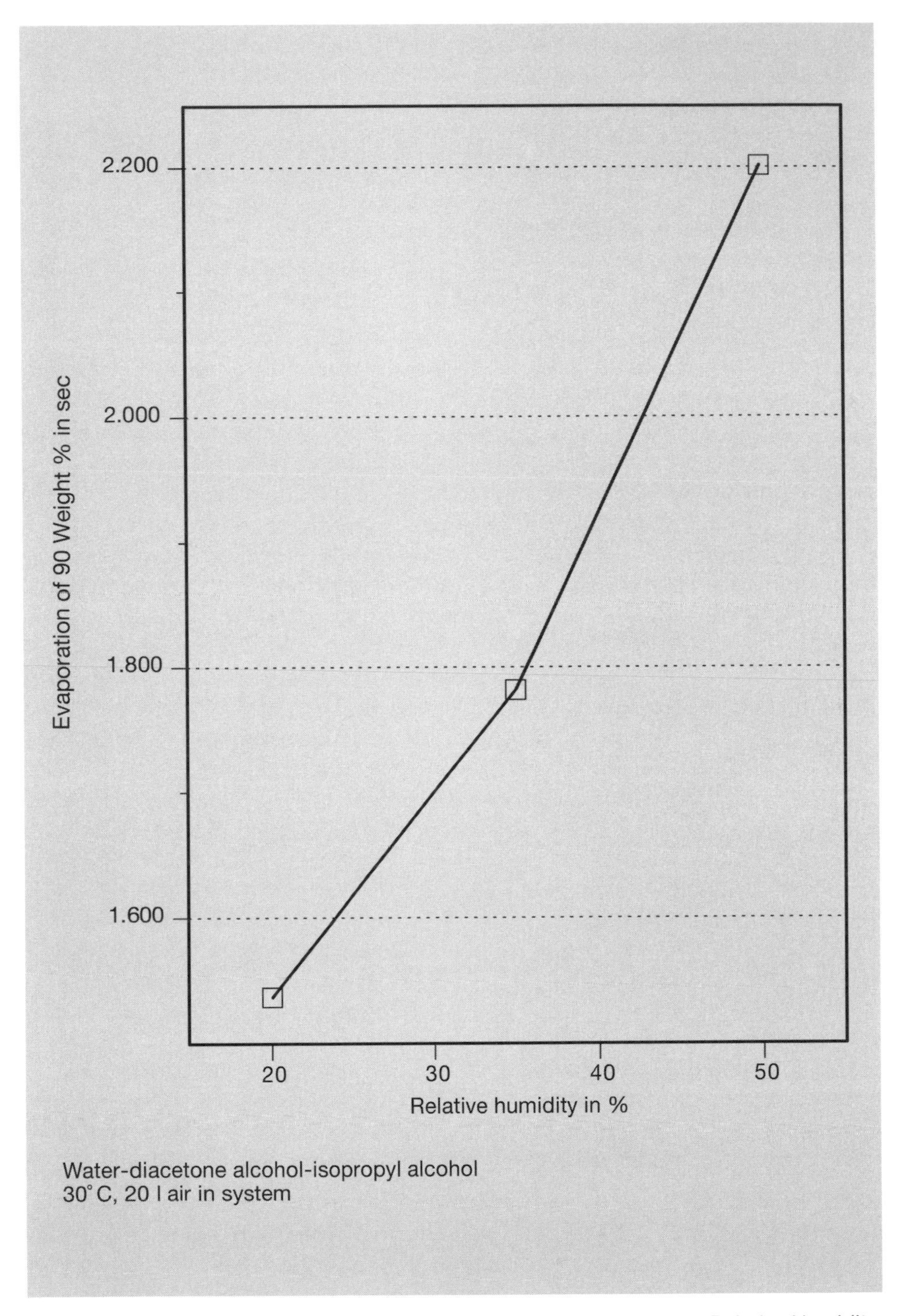

Figure 19 Evaporation of a Water-Solvent Mixture as a Function of the Relative Humidity

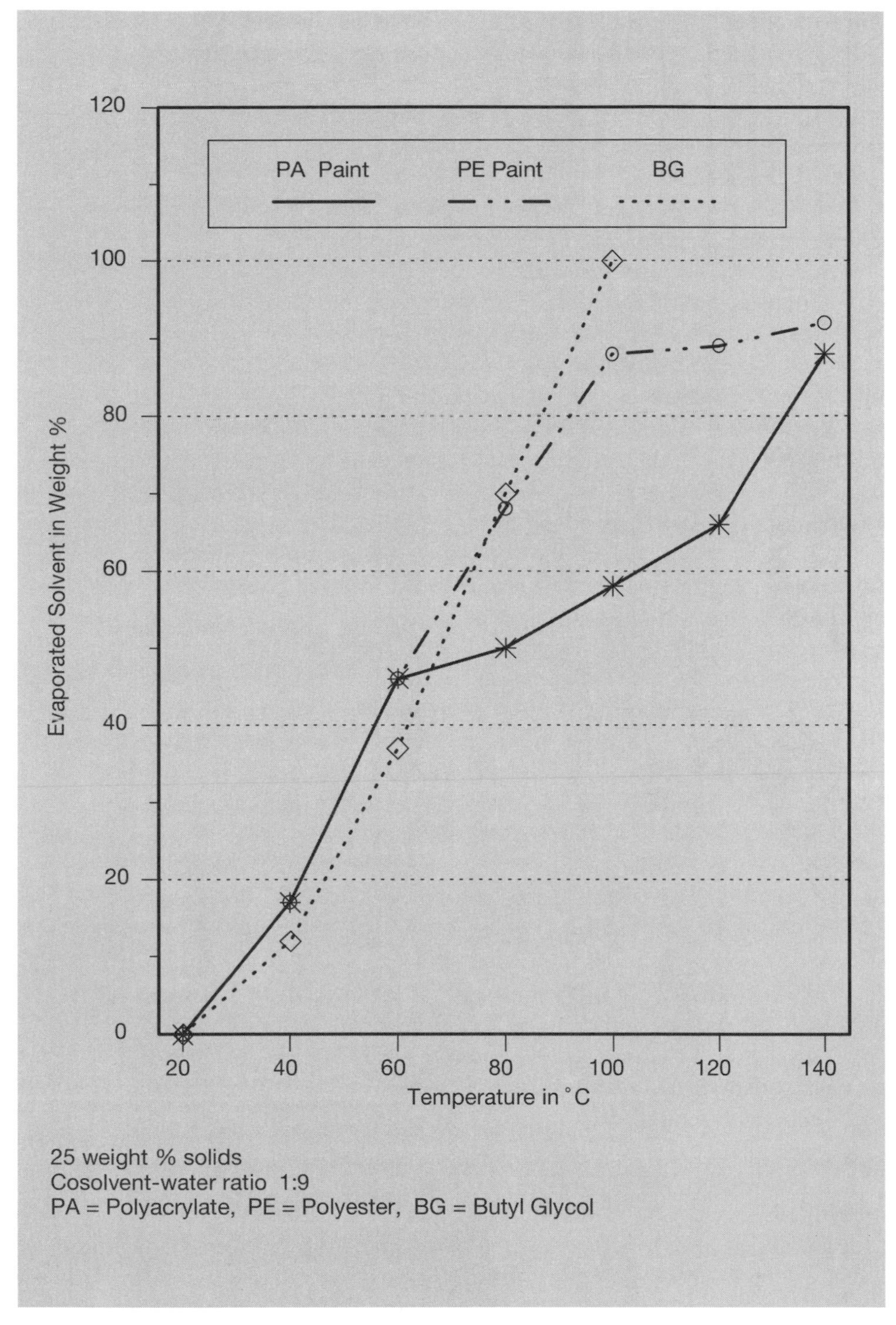

Figure 20 Evaporation of Water/Solvent from Water-Borne Coatings

These parameters are of decisive significance for the rheological behavior of the paint for example, as well as for body and flow, for tint and gloss development, and for wetting and film building. A certain solvent residue can also be necessary for contact with a subsequent coating. Humidity, temperature, and air circulation must therefore be adjusted for the prevailing coating conditions. One speaks here quite perceptually of the so-called applications window, e.g. value range of the above parameters, within which application and drying can proceed optimally [222, 223, 224, 225]. The demands on equipment are high, but by observing them excellent painting results are possible.

The customary cyclic procedures of industrial finishing aggravates the situation. Solvent-containing systems usually allow for quite short time exposure to air and steep heat-up curves with the forced air drying or baking processes. With aqueous paints, especially emulsion paints that dry from the surface, the water must be given sufficient opportunity to evaporate at temperatures below the boiling point [226]. Otherwise, especially with thicker coatings, explosive evaporation and thus destruction of the paint surface (blisters) occurs. Hot air and ir radiation are suited for accelerating drying, where the first primarily affects the paint surface (possibly negative) while ir radiation, depending on the wavelength can also penetrate to the deeper layers [227]. With the ir method, favored evaporation of water can be achieved (irradiation in the OH band region). Frequently a combination of both drying methods is advantageous.

In addition to forcing purely physical drying by energy application, the chemical cross-linking processes can also be accelerated. Included are the oxidative drying of alkyd resins, but also of styrene-butadiene dispersions and binders based on polybutadiene oils. Heating also promotes the polyaddition reactions of epoxy resin emulsions. Polycondensations of various types of resins and reactive polymer dispersions usually proceed significantly above 100°C under baking conditions, so that the water should be removed as completely as possible. Photopolymerizations are also of interest; there the hardening process consists of uv or electron beam induced polymerization [228]. The method is suited especially for the industrial coating of metal, plastics, paper and wood where high quality coatings in very short time frames without great heat introduction are required. In addition to these industrial applications and hardening techniques, uv cross-linking dispersions are today used in highly elastic facade and roof coatings. Here the surface cross-linking effects a reduction in tackiness and thus soiling.

To describe the influence of the type of auxiliary solvents or cosolvents on the drying behavior of water-borne coatings, a rate constant of volatility was formulated [222, 223]. This empirical parameter predicts how many grams of volatile components – influenced by the given solvent used -escape per minute and square meter from the coated surface at 25°C. The rate constant of volatility (g/min m^2) has the following values for the solvents investigated (table 18):

Table 18 Rate Constants of Volatility of Solvents

Solvent	Rate Constants of Volatility of Solvents g/min m^2 at 25 °C
Butyl Glycol	1.68
Propoxypropanol	4.72
n-Butyl Acetate	21.50
Isopropyl Alcohol	61.10
Methyl Ethyl Ketone	136

A quantitative investigation of the parameters on the drying of water-borne coatings shows that the relative humidity and the drying temperature are of special significance, while the air rate and variations in the water-solvent ratio make only a relatively small contribution to the drying behavior. Table 19 gives a quantitative comparison of the effect of parameters within the definite variation ranges that were investigated:

Table 19 Relative Effects of Parameters on the Evaporation Time of Water-Solvent Mixtures

Parameter	Range Investigated	Effect in %
Relative Humidity %	10–90	60.2
Temperature °C	20–40	20.2
Air Speed l/min	10–30	1.6
Added sec-Butanol %	0–100	4.5
Water Content %	60–100	under 0.1
Other		10.2

The concentration ratio of water to cosolvent should not change uncontrolled during the drying procedure. The relative humidity at which this change is zero is designated the critical relative humidity (CRH).

Table 20 Solvent Groups with Comparable Critical Relative Humidity (CRH)

Solvent	CRH	Rating
Methyl Ethyl Ketone	0 %	low
Isopropyl Alcohol	0 %	low
Propoxypropanol	24 %	moderate
Butyl Glycol	74 %	moderate
Ethyl Diglycol	93 %	high

If the air is dryer, a greater portion of the water evaporates and the solvent remains in the film. If the air is wetter, larger quantities of solvents volatilize and the water content of the drying film increases. Both effects can cause technical problems, as for example an undesired temporary softness of the film, the simulation of higher flexibility, the appearance of turbidity, surface haze, etc. The critical relative humidity depends on the type and quantity of cosolvent [219, 222, 229]. As an overview, solvents can be divided into three groups: low, moderate and high critical humidity. One should attempt to use such cosolvents that offer moderate critical humidities (CRH). Solvents are divided by groups in table 20.

15 Energy Balance for the Drying of Water-Borne Coatings

As a result of the higher enthalpy of vaporization compared with the usual paint solvents (see chapter 5), the evaporation of water from coatings requires a significantly higher energy usage. In view of its vapor pressure at room temperature and its relative volatility, water is a moderately to difficultly volatile solvent. The energy for the application and cross-linking of aqueous baking paints in comparison to other coatings systems with the same external conditions (same paint surface, drying at 180°C, 20 min) has been reported by various authors [230] (table 21).

Table 21 Energy Expenditure for the Application and Cross-Linking of Various Coating Systems

Paint System	Energy Usage in kcal/hr	Solids Content
Baking Paint, Conventional	9213	30 wt. %
Water-Borne Coating	6195	30 wt. %
Powder Coating	5305	
High-Solids Coating	6070	80 wt. %

It is apparent, that the energy usage is highest for solvent-containing paints, while the energy input required for water-borne coatings is comparable to high-solids coatings and is only slightly higher than for powder coatings. An advantage of the processing of water-borne coatings is the non-flammability of the systems. While for safety reasons, the quantity of heated circulating air must be very high in the case of solvent-containing coatings to keep the concentration of entrained volatile solvents below the explosive limit, the required circulating air for water-borne coatings can be lower, since here only water vapor, small quantities of auxiliary solvents and possibly condensation products are present and must be removed. This frequently represents a significant energy savings. In the case of the use of water-borne coatings in the automotive industry it was shown that water-borne coatings can be applied for similar energy costs as solvent-containing coatings [230]. For all these reasons, the advantages of water-borne coatings are valued highly so that these products are given a good probability of success in various application areas [11, 231, 232].

16 Applications for Water-Borne Coatings

The current state of water-borne coatings technology allows for completion of the transition from conventional coatings systems to aqueous products in more and more areas. This change is often not as problem-free as it may appear initially. In the simplest case, matching the applications procedure to the new materials properties is all that is required, as for example for corrosion protection systems in the trade or in the do-it-yourself area. The introduction of water-borne coatings in many industrial areas is a different matter. Here, the continued use of old processing techniques is usually not possible, or in the final analysis is too expensive. The processes must be conceptualized in detail and implemented from pretreatment, through the actual painting to the disposal of paint wastes. In such cases a high financial and personnel investment is compelled.

In the following, examples from important industrial and trade areas are discussed. Some aspects of current applications areas are described, in which the transition to aqueous coating systems is presently occurring. Reference is made to the original literature for a detailed treatment of specific applications areas. The area of paint coagulation and disposal or recycling of paint residues is discussed in chapter 17.4.

16.1 Industrial Coatings

The area of industrial coatings is extraordinarily varied. The technical requirements as well as the applications techniques encompass a broad spectrum. The easiest area to visualize is probably that of automotive paint lines. As an introduction, something will be said about the pretreatment (see also chapter 13.1) of above all, metallic surfaces. Since the requirements vary greatly, agreement with technical firms should always be obtained in other industrial coatings areas [233].

The same is true for the spray and design technology used [234]. Special consideration must be given to a high degree of paint transfer efficiency, since in principle, the problems of disposal and water-borne coatings coagulation do not differ from that of solvent-containing coatings. Wherever possible electrostatic procedures should be favored [235]. Table 22 gives an overview of the most important processes, their transfer efficiency and frequently required modifications (see next page please):

It should be evident that only high-efficiency transfer processes are capable of minimizing paint wastes. The constantly increasing disposal costs for paint coagulates has lead to various recovery concepts (see chapter 17.4).

Table 22 Spray Processes for Water-Borne Coatings

Process	Required Modifications	Transfer Efficiency
Pneumatic	Increase pressure	ca. 30%
Airless	Exclude air bubbles	ca. 40%
Electrostatic	Isolation, Atomization	60–99%

16.1.1 Automotive Coatings

The automotive industry is often a pioneer in the coatings area with respect to the introduction of new technology [232, 236, 237, 238, 239]. Automotive painting meets the corrosion-protection and decorative requirements to a high degree. The quantities of paint materials used in automotive factories are substantial, so that they are a matter of interest with respect to environmental discussions. In Germany about 5 million vehicles were produced in 1991. Thereby the paint usage in the automotive industry may have reached 100,000 metric tons, with more than 50,000 metric tons of volatile organic components. For Europe, about triple these quantities are projected for this applications area [225].

Passenger car painting encompasses, after pretreatment, the steps of electrocoat priming, application of filler, base coat and clear coat, which will be subsequently be considered more fully. The relationship between film thickness and solvent emissions is depicted in table 23 [212]:

Table 23 Solvent Emissions as a Function of the Paint Film Thickness

	Film Thickness (µm)	Solvent Emissions (g/m^2)
Cathodic ED Primer	20	3
Filler, Conventional	35	39
Base Coat, Conventional	15	54
Clear Coat, Conventional	35	25

Accordingly, metallic base coats, followed by filler and clear coat, allow the greatest reduction in solvent, when for example a change is made to low-emission aqueous coatings substances. The anticipated regulation for the reduction of organic emissions in Germany targets a maximum quantity of 30 to 40 g solvent per square meter of coated body surface, while in the remainder of the European Community 60 g/m^2 for metallic paints and for uni-paints will be accepted [240].

Since not only solvents contribute to environmental pollution in automotive body painting, initial comprehensive experiments regarding energy and material considerations were conducted [241]. These encompass

- Air (organic solvents, residual mists)
- Water (organic and inorganic constituents)
- Solid Waste (paints coagulates, hydroxide sludges)
- Energy (heat, CO_2, NO_x, SO_2)

Based on this one can anticipate the following developments in automotive manufacturing by the year 2000:

- improved CED primers (reduced baking temperature and solvent content)
- high-solids or aqueous fillers
- aqueous base coats
- aqueous or powder clear coats

16.1.1.1 Pretreatment Techniques (Phosphating)

Today's 6 to 10 year warranties against rusting by the automotive producers are based on optimal pretreatment (see chapter 13.1) in addition to electrodeposition priming. The surface is made more uniform and the paint adhesion is improved. Naturally these pretreatment techniques are used not only with motor vehicles, but generally in industrial coating where definite resistance properties are required [233].

What is the process in detail? The stamped body is cleaned and degreased [242]. The actual phosphating follows, which today is used especially on steel, galvanized steel and aluminum. Phosphating can be laid out as a spray, spray-flood, spray-dip or pure dip process. In automotive manufacturing, zinc phosphating has long supplanted iron phosphating. Further development of the process has led to so-called low-zinc phosphating, which is in part manganese modified. With this a zinc-treated dipped surface free of specks is possible which in addition has higher resistance. Spray-dip phosphating of automotive bodies encompasses the following steps [181]:

Spray degreasing – dip degreasing – spray rinsing
- dip rinsing + activation – phosphating
- spray rinsing – dip rinsing – pacifying with Chromium [III] and [VI] by the dip process
- DI water dipping – DI water rinsing

Phosphating times for the dip process lie between 3 and 7 minutes and 2 minutes for the spray process. The expenditure in pretreatment is substantially determined by the corrosion protection to be achieved. Careful coordination with producers is essential.

16.1.1.2 Electrodeposition Coating

Electrodeposition painting (see also chapter 13.2.6.) is today a fully-automated coatings process that contributes significantly to the life of an automobile. The paint material is electrochemically deposited on the automotive body and then hardened by baking.

The baths based on solvent coatings used previously have been supplanted by these aqueous systems since the mid-sixties due to the former's high hazards. While originally, binders based on maleinized oils, alkyd resins and epoxy resin esters were used, polybutadiene systems began increasingly to dominate at the beginning of the seventies. These afford very good corrosion protection, good coatability and bath stability. The process designated as anodic electrodeposition or AED is based on negatively charged polymer particles, which deposit on the positively charged body. Today polybutadiene systems again have increasing significance for automotive supply parts and for the industrial dip priming of metallic parts [243, 244, 245, 246, 247, 248]

In the eighties, based on developments in the USA after 1976, the so-called cathodic electrodeposition priming or CED gained dominance for passenger car manufacturing [33, 249, 250, 251]. Usually amine-modified epoxy resins, which are cross-linked with blocked isocyanates form the chemical basis. The primary advantage of cathodic coatings is their further-improved corrosion protection, even in thin films. Disadvantages, such as elevated baking temperatures and expensive facilities are today under control. While the original systems were designated as thin-film CED (12 – 20 µm), the next generation is called thick-film CED (to 35 µm). They are said to provide improved stone chip resistance and coverage of the steel roughness. The deposition process on galvanized steel has also been optimized. With the dip priming designated as 2C-CED [212], a cross-linkable special resin for pigment dispersion and a catalyst is introduced in addition to resin, blocked isocyanate and cross-linkable plasticizer. Very low solvent contents of 1 to 2% are thereby achieved.

Currently the further reduction of the solvent content to under 1%, lowering baking temperatures to 150 or 125 °C and improved stone chip resistance at lower temperatures are being worked upon. This new CED should by formulated to be as lead-free as possible and require practically no rejection of ultrafiltrate [252]. Products newly brought to market [253] come close to meeting these requirements. They offer in addition, improved coverage of grinding marks and reduced baking losses in the oven [211].

16.1.1.3 Aqueous Fillers

The filler, as also the sprayed primer, provides for a further uniformity of the surface. Good corrosion protection is expected since it must cover worn areas of the dip priming. Additionally, it must have adequate stone chip resistance, body and can be polished [254]. It serves as an adhesive bridge between

CED primer and top coat system [242]. Aqueous fillers were the second aqueous component in automotive coating after electrodeposition paints. Introduction began in 1981. Polyester/melamine and polyurethanes are known as binder systems [240]. The polyurethane primers are superior in their applications properties but are expensive [212]. While the systems mentioned require about 5% cosolvent, combinations of acidic resins with dispersions contain only about 1% amine as neutralizing agent [14]. The following representation shows some properties of aqueous fillers:

Processing Viscosity	s	30 - 45/50
Processing Solids	wt.-%	52 - 60
Volatile Organic Components	wt.-%	5-7 - 1
Baking Temperature	°C	160 - 170
Baking Time	min	15-20 - 25
Film Thickness	µm	35

The fillers have a crosshatch of 0, Erichsen impact of over 5 mm and are resistant to salt spray for at least 240 hours on bare steel. Application can proceed by all customary procedures; as a rule fully-automatic electrostatic spraying with high-rotation bells are used. Only the requirements of the electrostatic spraying technique, as they are known today, must be fulfilled.

At this point it should be mentioned, that the usual seals (welding seams, edges) and the underbody sealing are simultaneously hardened with the primer. Application of these materials usually takes place earlier.

16.1.1.4 Aqueous Base Coats

The aqueous base coat is the fundamental component of metallic painting. It is applied after polishing the filler, which today is done only on a spot basis with wet polishing, and after cleaning with the brushing machine. Conventional base coats are the principle cause of solvent emissions in automotive painting. The paints originally contained only 13-15% solids, now 18-20% are possible. Aqueous base coats were and are the focus of research in the paint industry. The first transition in practice occurred in the second half of the eighties. Problems to be solved were for example the applications techniques, color tint availability and repair friendliness [242]. Polyester, polyacrylic and polyurethane resins serve as the polymer basis [55]. At present, the aqueous base coats contain about 15% organic solvent. Coating with clear coat is not restricted, the same goes for weathering resistance [212]. Here some typical properties [225]:

Processing Viscosity	s	27
Processing Solids	wt.-%	20
Volatile Organic Components	wt.-%	15
Film Thickness after Intermediate Drying with Infrared/Hot Air	µm	12-15

An important processing aspect is the kinetics of drying, since – as with solvent paints – only certain quantities (maximum 10% solvent or water) can remain when the clear coat is applied. The influence of temperature, humidity, air velocity etc. must be known precisely, to attain optimal results [218] (see also chapter 14. and 15.). Pigmentation has been investigated in equal depth [139]. For clarification, the simplified painting process, as used in the Opel works [242], is presented:

- Cleaning with brushes
- water-borne coating exterior, electrostatic
- water-borne coating interior, manual, air-atomized
- water-borne coating exterior, air-atomized
- Intermediate drying with infrared and hot air
- Clear coat interior and subsequently exterior

16.1.1.5 Automotive Clear Coats

Aqueous clear coats are presumably the last step in the introduction of water-borne coatings technology in automobile manufacturing. They are being worked on intensively in various places. The first practical applications are about to be expected, positive tests having been obtained in large-scale facilities. The weathering resistance of the aqueous paint systems is rather better than of the conventional. The rheological properties remain to be improved [240].

Here some properties of a successfully tested clear coat based on an acrylic resin with 43% solids and 14% cosolvent content [255, 256]:

Processing Viscosity	s	28
Processing Solids	wt.-%	43
Volatile Organic Components		
– Solvents	wt.-%	14
– Amine	wt.-%	0.6
Baking Conditions		
– Predrying		7 min at 80°C
– Circulating Air Oven		25 min/140°C

Solvent emissions are reduced about 50% compared with low-solvent assembly line paints. While the tests regarding repair capability and car wash stability are positive, work is being done to optimize electrostatic application.

In addition to aqueous clear coats, intensive work is in progress with high-solids clear coats and powder coatings [212]. Powder coatings are given a good chance of success for the distant future since, in combination with the use of aqueous coatings up to the base coats, the lowest emission value of all are possible [30 to 40 g/m²). For illustration , the solvent emissions per passenger car, as present at Opel [242], in table 24:

Table 24 Solvent Emissions per Passenger Car

	kg Solvent per Mid-Size Passenger Car	
	Conventional	Aqueous
Cathodic ED		0.2
Filler	1.0	0.2
Base Coat	6.0	1.2
Clear Coat	2.7	n/a
Total	ca. 10	1.6–2.5

With a powder clear coat, the emissions could thus be reduced to nearly 1.6 kg per passenger car with the current state of aqueous technology. With the use of an aqueous clear coat less than 2.5 kg emissions should result in the total system.

16.1.2 Metal Coatings

Coating with aqueous paint materials in general industrial painting is much less standardized than in automotive painting. Especially in metal coating, many paint systems of differing binder basis compete with each other [257, 258], since on the one hand, a multitude of requirements exist and in addition the suppliers of coatings raw materials strive to offer all the traditionally-introduced raw materials also in water-borne coatings quality. In general water-borne coatings have achieved the same level of quality as solvent coatings. Metal coatings are broadly divided into the areas:

- oven drying for interior and exterior application,
- air or forced-air drying for interior and exterior.

In addition to primers, these are mostly a matter of one-coat paints.

The type of pretreatment of the metal is determined by the required corrosion protection (see chapter 16.1.1.1.). It must be suited for the given object as must be the painting. Here close cooperation between the user, paint supplier and equipment manufacturer is required. In the following chapters there are some examples for clarification.

16.1.2.1 Aqueous Baking Primers

A range of suitable binder materials exists for aqueous baking primers, often based on hydrosols or hydrogels, but also emulsions. Cross-linking is usually with melamine resins as in conventional paint chemistry. Interesting combinations of aqueous primers and powder top coats have been developed [259]. Substantially higher resistance has been obtained on steel, galvanized steel and aluminum, where the necessity of degreasing or phosphating is determined by the individual case. Some properties are collected in table 25. As part of a total paint system, better properties have naturally been achieved. Water-reducible baking primers are suitable for radiators, steel furniture, exterior components, machines, etc.

Table 25 Properties of Aqueous Baking Primers

	Aqueous Priming on Steel		Aluminium
	Degreased	Phosphated	
Film Thickness µm	40	35	30
Baking Temperature °C	180	180	180
Baking Time min	20	20	20
Cross Hatch	0	0	0
Erichsen Cupping mm	>5	>5	>5
Pendulum Hardness König sec	100	100	100
Salt Spray hrs	300	500	1000
Tropics Test	300	500	1000

16.1.2.2 Baking Top Coats

One-coat paints have been used successfully with lights [14], drums [260], metal top coats [261, 262, 263, 264], bicycles [265] and many other metal parts.

Table 26 Properties of an Aqueous Structural Baking Coating

Binder	Acid Polyester / Melamine Resin 80/20
Solids, wt. %	45
Organic Solvent, wt. %	9
Application	Electrostatic Spraying-Pressurized Air *Base Paint:* 4 bar, 20 min Evaporation (Air 40°C, filtered) *Structural Paint:* 2.5 bar, 20 min Evaporation, 10 min 80–120°C, 15 min 140°C
Film Thickness, µm	60–90
Pencil Hardness (Gloss Coat)	H–2H
Erichsen Cupping, mm	5–6
Adhesion, Bonder 1041, Cross-Hatch	0–1
Resistance	Alcohol, Gasoline, Detergent

Usually these are a matter of combinations of binders and cross-linking resins. The paints contain between 5 and 15% auxiliary solvent. An interesting new development is represented by the 3C-water-borne coatings. They are completely solvent and amine free, but demand a special technique for the mixing of epoxy resin, hardener and water directly before use. Shock absorbers are an example of their application [266, 267]. Disposal facilities are completely unnecessary. Table 26 shows the properties of a structural coating for office furniture, which has proven out well for many years in practice.

16.1.2.3 Coil-Coatings Systems

In coil coating also, there are now examples in practice of coating with aqueous paints. Thus top coats based on saturated polyesters and melamine resins are mentioned. In Scandinavia, top coats based on pure acrylate emulsions are in use. They can be applied on conventional or water-based primers in the same manner. The exterior stability is good as anticipated, but the optical properties are not yet optimal due to the poor flow of the emulsions [46, 268]. Water-reducible coatings have also been introduced in the packaging and can coatings sectors [269, 270].

16.1.2.4 Air and Forced-Drying Primers and Paints

In addition to corrosion protection (see chapter 16.2.), these coatings materials are especially used for large parts. Examples are rolling stock, construction machinery, agricultural equipment and busses [271, 272, 273]. They should be capable of being applied with as few difficulties as possible using current procedures (flow cups, pressure vessels, airmix, airless) and on all customary substrates (cast iron, degreased, phosphated and sand-blasted steel, non-ferrous metals). Priming is specially geared toward spray priming, dip priming, welding primers and thin-film primers [274). The oxidatively as well as physically drying primers have up to 8% solvent content [14], newer developments lie around 2% [274]. Higher demands with respect to durability apply for exterior top coats. Especially demanding are coatings for rolling stock, whose chassis must last over 20 years. Top coat refurbishing is undertaken after 8–10 years. One- and two-component variants have been tried as primers, acrylate emulsions and alkyd resins are under trial as top coats [275, 276].

16.1.3 Plastics Coatings

Plastics, as versatile raw materials for articles, play an increasing role as substrates to be coated. The requirements are extraordinarily numerous. They reach from electrical to recreational equipment and automotive manufacturing [277]. There is a whole range of plastics to be coated. Named as examples are: ABS-rubber, epoxy resin molding compounds, EPDM copolymerizates, polyamides, polybutylene terephthalates, polycarbonates, polyacrylates, polyphenylene oxide, polystyrene, polyurethanes, PVC, phenolic molding compounds, and polyolefins such as polyethylene or polypropylene [53,278,279]. Each raw material has its own definite properties, so that tailoring the coating material to the plastic surface is essential. The goals of plastics coating can be:

- attaining definite color tints, degree of gloss, metallic effects
- matching to the piece to be completed
- covering or eliminating surface defects
- improving scratch resistance, chemical stability, weathering resistance.

Care must be taken during coating that the plastic is not impacted negatively, as for example changing the rigidity, the cold flexibility, the elongation behavior or by softening, crack and blister formation. A variety of binders are available for plastics coatings, where recently polyurethane dispersions are playing an increasing role [53].

In the following the situation in the automotive supply industry is discussed in more detail. The varied quality requirements lead one to expect that the solutions from this area will be rapidly transferred to the total industry. Plastics are indispensable in automotive manufacturing above all because of their capability to be shaped and their elasticity (bumpers). To be sure, requirements for recycling capability present new tasks, so that "pure" polymerizates are increasingly gaining favor over polymer combinations. The build-up of the coating often corresponds to automotive body painting e.g. primer, base coat and clear coat. For some materials, suitable aqueous primers and base coats are already available or in the process of introduction. Clear coats are still based on solvents. Aqueous adhesion promoters are in trial for nonpolar polyolefins such as polypropylene blends [191]. Polycarbonate bumpers are being coated with aqueous paints on line. The build-up consists of aqueous priming, two layers of base coat, and two-pack polyurethane clear coat (solvent-based, drying at 90°C) [280]. Aqueous primers are already serviceable on line for polyurethane parts (PU-RIM), base coats are under development [281]. Electrodeposition coating is now also available for plastics [282]. Anodic systems, precipitated on a galvanized substrate and baked at 70–80°C, are used.

16.1.4 Industrial Wood Coatings

Industrial wood coating will be discussed using the examples of furniture and window painting. Until about 10 years ago, it was customary to install unpainted windows in buildings before plastering was done. The effect of water before painting led to inadequate durability which led to the decrease in market share of wood windows. Also the customary use of decorative paints or thin-layer glazes in the eighties, both applied to glazed and impregnated windows, led to a variety of shortcomings [283]. Sealed with plastic caulks, swelling, shrinkage and separation occurred. Today most wood windows are painted industrially before glazing and installed complete. In combination with elastic caulks and joint compounds, the market share of wood windows in Germany again increased to 40%. Shifting the painting operation from the construction site to the factory simplified the use of aqueous coating materials, since wood as a so-called "living" material presents some special requirements. When considering coatings materials, one must distinguish between glazes and covering coatings [283]. Glazing consists of

- water-reducible dip impregnation with protection against algae and wood-damaging fungus and
- water-reducible thick-film glazing with 250 µm wet film thickness (80 µm dry) by airless application.

For covering paints is used consisting of

- synthetic resin-dip primer equipped against algae and wood-damaging fungus
- acrylate spray filler with 150 µm minimum wet film thickness and
- acrylate top coat with 150 µm minimum wet film thickness.

The dry film thickness of the total water-borne coating should not lie below 100 µm to enable guarantees of three years.

In addition to the paint systems mentioned there are also water-reducible 2-pack materials which can be used in fully automated paint lines [284, 285].

The coating of furniture, in contrast to windows, is much more subject to fashion trends. Aspects such as flexibility in color and texture and object-determined finishing and painting take lasting effect. In addition to traditional materials such as nitrocellulose-coatings [287], acid-hardening coatings, polyurethane and phenolic coatings, radiation-curing (uv and electron beam) and water-reducible systems are being introduced. In part, uv curing and the use of water are being combined, as for example in coating chairs with electrostatic application [286] or in the manufacture of bedroom and living room furniture [288]. In recent years the following were introduced for the latter:

- uv aqueous primers
- uv roll primers
- uv top coats
- aqueous stains and equalizing stains
- aqueous fillers

The selection of paint materials depends on the desired color and porosity (open-pored, half-closed or through-colored). Where possible roll processes are used, to be sure fronts often need to be poured. Spray application has always required special care, since the coatings material-dependent coagulates adversely affect the environment. Here too detailed investigations have shown the advantages of uv water-borne coatings [288].

16.2 Corrosion-Protecting and Repair Coatings

Naturally the aqueous paints for corrosion protection and repair cannot be clearly separated from the systems already described (chapter 16.1.2.) under the concept of metal coating. At this point, it will be less a matter of industrial paint lines, but about materials that are used externally in the trade, for example by corrosion protection firms. Working conditions are characterized by usually having to paint newly installed pretreated parts (galvanized or primed) or having to paint in facilities in use. As a rule only spraying and brushing are feasible. Adverse weather conditions such as cold and rain make the work additionally difficult. For this reason, the introduction of water-borne coatings here is only at a beginning. Since costs are strongly determined by the labor used and less by the material, and since warranties are usually required, the tendency is toward proven coatings systems based on solvents. Such systems often contain alkyd resins, epoxy resins or their fatty acid esters, VC copolymers and polyurethanes or polyisocyanates as the binder material. The solvents contained in the paint simplify substrate wetting and enable a certain adhesion to inadequately cleansed (usually blasted) surfaces. The demonstrated durability of conventional coatings is also required of water-borne coatings.

16.2.1 Primers

Corrosion-protection primers are taking over the active protective function of the coatings system. In addition they must have good adhesion to the substrate as well as to the intercoat or top coat. Aqueous primers are also required to make do without carcinogenic chromate pigments. The corresponding phosphates (zinc phosphate) must therefore be used. Flash rust can be prevented with suitable additives (amines) [26]. The binder basis for the primer is determined by the substrate. On steel, for example, alkyd resins or epoxide ester resins or their combination can find application. Commercial styrene-acrylate and pure acrylate dispersions are usually not suited, since their emulsifier content does not allow good corrosion protection. Producers are seeking to gain a foothold in the market with so-called secondary dispersions, i.e. solvent polymerizates, from which the solvent is removed by distillation after the addition of water. Styrene-butadiene emulsions have proven out best [62, 63]. They possess an elastomeric character and can be oxidatively cross-linked by the addition of dryers. Good adhesion and little tendency to blister are achieved on steel. They are also used for stone chip protection in automotive manufacturing. Spraying is the preferred applications technique.

Fatty acid-modified binders such as alkyd resins are suitable for galvanized substrates only in exceptional circumstances (e.g. in combinations). Styrene-butadiene emulsions also do not adhere satisfactorily in all cases. Polybutadienes, commercially available as emulsions and also as solutions, have proven out well here. These binders bind to steel, nonferrous metals, wood and concrete. They brush well and are suited for manually-cleaned steel. Table 27 shows a comparison of the properties of a styrene-butadiene emulsions and of a modified polybutadiene (water-soluble) in primers. Both systems have been exposed to an industrial atmosphere for many years on a tank with and without top coat.

For so-called heavy corrosion protection, two-component epoxide resin systems are used among others. These are also available as emulsions [289]. Higher molecular epoxide resins with a copolymerized emulsifier are the basis. The hardener should be a modified amine. Good resistance values are obtained in combination with zinc phosphate as corrosion-protection pigment if adequate temperatures (at least 15 °C) and wet film thickness are provided. Epoxide resin emulsions presently available on the market still contain solvent (e.g. methoxypropanol, below 10%). Fully solvent-free products are however already under development.

Table 27 Properties of Corrosion-Protection Primers

	Styrene-Butadiene Emulsion	Modified Polybutadiene
Solids	57 wt. %	62 wt. %
Solvent Content	2.8 wt. %	8.8 wt. %
Substrate Steel	+	+
Substrate Zinc	–	+
Rust Removal Manual	limited	+
Rust Removal Blasting	+	+
Brushing	limited	+
Salt Spray Test	240 hr	1000 hr
Ambient Weathering (5 years)	No Rust, Moderate Chalking	No Rust, Moderate Chalking

Aqueous coatings are increasingly being evaluated and accepted by the national railroad companies [272]. A large number of systems was tested on hot-galvanized as well as blasted steel surfaces in the laboratory and in the open. A selection of positively-evaluated systems has been applied to bridges and electrical masts. Pure water-borne coatings systems displayed spot rusting when they were applied to blasted surfaces. A system consisting of epoxy-zinc dust primer and two aqueous top coats has until now passed the four-year use test. Hot-galvanized substrates have proven to be less problematical. Numerous water-borne coatings systems were equal to PVC, epoxy and PU coatings. The results indicate that careful trials are necessary in each case to avoid expensive re-working. As a rule, complete judgments about definite corrosion-protection coatings are not possible. The results must be evaluated based on the substrate and the applications conditions.

The entire area of corrosion protection is presented in the German Standard DIN 55 928. There the currently established state of water-borne coatings technology is taken up [290].

16.2.2 Intermediate Coats and Top Coats

While intermediate coats above all strengthen the barrier effect of the paint and increase the total film thickness, top coats have the assignment of guaranteeing resistance to light, weathering and mechanical effects. Although as a rule, aqueous top coats are practical over conventional primers, wherever possible the total system should be laid out on an aqueous basis. In each case matching the applications conditions and the buildup of the individual layers is essential. It is safest to work with specifications that are recommended and evaluated in practice by the paint producer. The composition of water-borne coatings differs only little from the solvent coatings. Depending on requirements, a range of proven air-drying binders such as alkyd resins, urethane resins and acrylate emulsions are available. The latter especially are distinguished by low chalking in industrial climates [26]. In addition to pure acrylate emulsions, cost-effective styrene-acrylate emulsions are in use [117] which attain an even further improved range of properties.

With respect to the climactic conditions for application of aqueous corrosion-protection systems, certain limits must be observed. Temperatures below 10°C and relative humidities above 80% are often critical. Also adequate air circulation must be guaranteed (see chapter 14). Before renovation of old coatings, it must be established that the wet adhesion of the water-borne coating suffices.

Under favorable conditions, it is possible with rapid drying water-borne-coatings to complete a total buildup of 250 µm (three layer) in one day.

16.3 Painter and Do-It-Yourself Areas

The statements made for other applications often apply in similar fashion to the painter and do-it-yourself areas. Of special interest are coatings for wood, metals, plastics and minerals. The danger from solvents, toxic contents or byproducts from the coatings materials is immediately present here, since in the best case personal protective measures such as gloves and masks are possible, which to be sure might be avoided for various reasons or even declined. On the other hand, the trade applications techniques and the decades-long familiarity with the traditional paint materials require a precise adaption to the alternative paint materials

16.3.1 Wall Paints for Interior and Exterior Areas

Today paints for building protection and interior walls are overwhelmingly produced based on polymer dispersions. Over 50 years of development have given aqueous paints derived from polyvinyl acetate a dominant market share. In the case of facade coatings, development has proceeded in the direction of saponification-resistant copolymer dispersions based on various monomers. Pure inorganic 2-component silicate systems are also known. One-component silicate paints contain polymer dispersions for stabilization [61]. In addition, these serve to improve the moisture resistance, chalking behavior and adhesion. The most modern facade coatings at present are based on combinations of silicon resins and polymer dispersions in varying ratios [59]. Expensive silicone emulsions provide better water vapor permeability, rain protection, and chalking and color stability. The currently still dominant silicon resin emulsions contain ca. 15% solvent, which results in ca. 3% solvent content in the finished paint. Until now this has restricted interior use [60]. Newer developments avoid volatile organic components, in that the silicone resins in the aqueous emulsion are first converted to the polysiloxane. To be sure, to keep the entire paint system solvent-free, the primer, which contains only 10% solids, must also be converted to an aqueous base. This has succeeded with new silicon micro-emulsions [60]. The dispersion paints for interior areas may be presumed to be familiar due to their long tradition. For special requirements emulsions are also used. Thus especially epoxide resin emulsions can be used to fulfill high demands for chemical resistance or cleaning capability (kitchens, hospitals). Similar to the old oil-based paints, are alkyd resin emulsions which are also suited to nicotine-resistant paints in restaurants.

16.3.2 Wood Primers, Glazes and Paints

The area of wood coatings still strongly bears the imprint of experience with alkyd resin paints. As in the case of materials for trade metal coatings, advantages of water-borne coatings such as quick drying, low odor and good color tint stability are accepted. Problems such as low brushing resistance, poor flow, thermoplasticity and difficult tool cleaning cause skepticism on the part of the painters. Restrictions with respect to humidity and temperature lead to further resentment [291].

Never the less the development to environment-friendly systems is also in full gear with wood coatings. In addition to aqueous coatings, high-solids systems will also play a significant role. Aqueous impregnation of wood can be done with substance-containing materials based on alkyds resins [292], acrylic dispersion, liquid polybutadienes, etc. Glazes and paints are usually formulated with acrylates and alkyds. Often binder combinations or specially-formulated resins are used to simultaneously exploit the advantages of emulsions and water-soluble resins. Newer gloss emulsions should make possible aqueous alkyd resin-like paints with high gloss, good flow, good blocking resistance and adequate wet and dry adhesion also on old alkyd coatings. These are usually acrylate-based binders used with special associative thickeners.

16.3.3 Floor and Concrete Coatings

A range of water-based paints is now being offered for floors and concrete coating. Usually the same binders are used as with solvent-containing paints to attain special requirements. Frequently, products based on acrylic emulsions, alkyd resins or dispersions of chlorine-containing polymers suffice. If higher requirements with respect to rubbing and chemical resistance are required, two-component epoxy systems must be called upon [26]. In the future aqueous two-pack polyurethane materials should also become interesting for this area.

16.4 Road Marking Paints

Currently the market for road marking paints is still dominated by coatings materials based on VC copolymers and styrene acrylates. Both are applied today with at least 75% solids, i.e. high-solids.

Various aqueous paints are presently at the testing stage. Positive evaluations are being achieved by above all styrene-acrylate dispersions in corresponding formulations. Also in the future it will be difficult to obtain markings that can be driven on in the shortest time possible at low temperatures or at high humidity. Conceivable is the use of aqueous road marking paints in the summer and high-solids in the transitional seasons. To be sure getting the marking firms accustomed to the changed properties of aqueous paints (quick drying) appears to be difficult. The applications equipment must be suited for aqueous materials.

16.5 Paper Coatings and Printing Inks

Printing inks differ from paints primarily in their significantly lower film strength and the printing carrier used. The binder materials must often display above-average wetting properties to attain optimal color strengths in combination with high-grade pigments and colorants.

There are a variety of printing processes, where each presents special demands on the build-up of the printing ink. In addition to the desired increase in printing speed, the replacement of organic solvents naturally plays a large role within the framework of increasing environmental protection demands and health protection. Radiation-curing systems and aqueous printing inks are the materials of choice [293, 294] where aqueous radiation-curing systems are also being developed.

To be noted are the varied substrates, which can be both absorbing and non-absorbing, as for example paper, cardboard, cellulose, aluminum, polyolefins, nylon, polyester and PVC [295].

Aqueous flexo and intaglio colors have long been used on paper and paper board. To be sure, these are not solvent free, since mixtures of water and alcohols are used. New developments eliminate n-propanol, isopropanol and glycols and suffice with a maximum of 5% ethanol [296]. The alcohol reduces the surface tension of the water considerably and thereby enables better wetting and adhesion on differing print carriers.

Binders used up to now are acrylic emulsions, acidic maleic resins and polyurethane dispersions. Development increasingly also includes non-absorbing substrates and other printing processes.

17 Regulatory Considerations

17.1 Occupational Hygiene for Manufacturing and Processing Water-Borne Coatings

The Occupational Safety and Health Act of 1970 empowered the Department of Labor to establish occupational safety and health standards and established the Occupational Health and Safety Administration (OSHA). The act requires employers with eleven or more employees to prepare and maintain records of each recordable accident or illness. The Hazard Communication Standard establishes uniform requirements concerning employee access to information about hazardous materials in the work place. It requires chemical manufacturers to determine the hazards of their products and to disseminate that information by means of Material Safety Data Sheets (MSDS) and product labels. Employers are required to prepare a written hazards communication program including an inventory of all chemicals, the availability of safety information and employee training. Many of the provisions of OSHA have a direct impact on the coatings industry.

In the production of paints, resin solutions, solvents, pigments and fillers above all, act on the human organism as a result of inhalation or by contact with the skin. In the production of water-borne coatings, the use of hydrophobic, oleophilic products is greatly restricted, so that a fat-dissolving action, which can lead to skin damage by redness or inflammation, is reduced. At the same time, soiling of the skin and naturally also clothing can be eliminated by simple cleaning with soap and water. In spite of this, the handling of individual components can result in differing product-specific damage as known from conventional paints. Care is especially required with the addition of amines, since these frequently are more toxic than the solvents. The solvents used must naturally be handled with the same health measures, as are known to paint technicians. Of advantage is the fact that the quantities of organic products processed with water-borne coatings is definitely lower than with solvent-containing systems. Maintaining threshold limit values is thus significantly easier. Also storage and the safety precautions associated with storage are less critical.

The user of aqueous coatings will ascertain that water-reducible systems frequently are no longer designated as harmful to health and under prevailing regulations require fewer risk and danger statements and health advisories than conventional coatings. The reason for this is that the coatings designation is primarily based on the solvent content which can now be so low that the need for designation is obviated. To be sure, this must not lead to the indiscriminate use of water-reducible coatings in ignorance of the hazards that may still be present. Mists are created by spraying which can be inhaled with inadequate ventilation. These mists collect in the lungs and can cause agglutinate. That these agglutinations in the human organism are not favorable to health, in view of the complex composition of a coating, should be evident to everyone. The water-soluble coatings constituents also enter the organism by absorption through the skin. Avoidable wetting of the skin with the coating should therefore be prevented. In spite of the good occupational hygiene properties of water-reducible coatings, here too the wearing of protective masks when spraying and gloves when filling and diluting is to be recommended.

The Clean Air Act Amendments (CAAA) of 1990 set up a comprehensive system covering permitting, non-attainment areas, air toxics and enforcement. The act requires each state to develop a State Implementation Plan (SIP) to ensure compliance. The 1990 law greatly expands the original Clean Air Act of 1977. A major change is the imposition of criminal penalties for certain violations.

In Scandinavia, the so-called painting occupational hygiene air requirement (POHAR) is used to evaluate volatile pollutants. This value (PWA = Paint Technology Work Hygiene Air Requirement) gives the amount of cubic meters of air, that is necessary to dilute the emissions from one liter of coating to where the established occupational hygiene limits are achieved. Table 28 shows how the solids content and amount of solvent can affect the characterization and designation of coatings in Scandinavia. Amines added to water-borne coatings must be taken into account in the evaluation.

Table 28 Occupational Hygiene Air Requirement (POHAR) (PWA values) for Coating Systems

Coating Type	Solids Content (wt. %)	Solvent	PWA m^3 air/l
Solvent Coating	60	Xylene / Butyl Acetate 1/1	649
Solvent Coating (High-Solids)	80	Xylene / Butyl Acetate 1/1	324
Water-Borne Coating	50	5% Butanol	329

17.2 Flammability of Water-Borne Coatings

In contrast with organic solvents, the main solvent in water-borne coatings, namely water, has no flash point. Since however no pure water, but almost always a water-solvent mixture, is used, the question of to what degree the addition of water to solvents influences the flash point is of interest [297, 298, 299]. Table 29 shows a compilation of some experimental results. Water-isopropanol alcohol mixtures 60/40 are still flammable, a corresponding 70/30 mixture at room temperature in contrast no longer is.

The flammability of water-borne coatings depends on the quantity of solvent introduced and its flash point. With the use of butanol and butyl glycol as cosolvents in quantities up to 15 wt.% of the total paint, the flammable designation is as a rule no longer necessary. This should result in lowered insurance premiums.

Table 29 Flash Points of Solvent-Water Mixtures

Solvent or Solvent Mixture	Flash Point °C
Isopropyl Alcohol	+12
Isopropyl Alcohol/Water 50/50	+18
Isopropyl Alcohol/Water 40/60	> 21
Isopropanol/Water 30/70	Not Flammable
n-Propanol	+25
n-Propanol/Water 50/50	+34
n-Propanol/Water 40/60	Not Flammable
Butyl Glycol	+60
Butyl Glycol/Water 90/10	> 65

Table 30 Flash Points of Water-Solvent Mixtures

	Solvent/Water Ratio			
	1:0	9:1	1:1	1:9
	Flash Point (°C)			
Ethanol	10	19	25	51
Isopropyl Alcohol	12	19	24	41
Ethyl Glycol	46	57	72	–
Methyl Ethyl Keton	–9	–7	–6	5–9
Butyl Glycol	60	> 65	–	–

17.3 Emissions with the Processing of Water-Borne Coatings

The quantity of organic emissions from various coatings systems can be seen in table 31. The emissions of organic products from water-borne coatings is strongly determined by the quantity of organic auxiliary solvents and amines (see also [300]).

Table 31 Organic Emissions from Coatings

Coating System	Organic Emissions (Weight Percent Based on Solids)
Liquid Coating Conventional (50% SC)	100
Liquid Coating Medium-Solids (60% SC)	67
Liquid Coating 1-C High-Solids (70% SC)	43
Liquid Coating 2-C High-Solids (80% SC)	25
Water-Soluble	5–25
Aqueous Dispersion	3
Powder Coatings	0.1–4

Today the content of organic solvents in modern water-borne coatings lies between 2 and 12% in ready-to-use form. The emission of volatile organic products from the practice of preparing coatings depends on the film thickness, the thickness of the dry film and the ratio of the paint solids to the quantity of solvents including amines and byproducts.

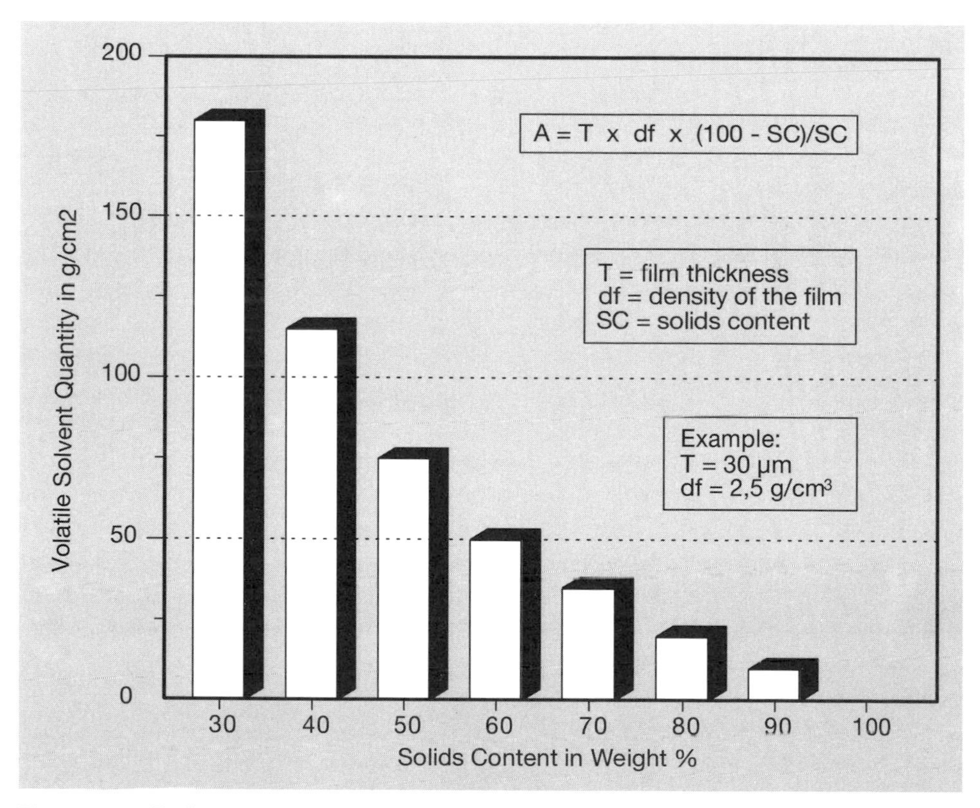

Figure 21 Emissions from Water-Borne Coatings

As a rule water-borne coatings have a solids content of ca. 50 weight percent in the ready-to-use state. With a content of 2 to 12% volatile organic products, they are thus comparable with low-solvent high-solids coatings with a solids content between 75 and 90 weight percent. As shown in figure 21, the emission protection effect of most water-borne coatings in use today are also in this range; they are thus to be viewed as more environmentally friendly than most high-solids coatings in practice for which the solids content seldom exceeds 70 weight percent.

In the coatings area, maintaining and remaining under the limit values specified by pertinent regulations serves as a criterium for evaluating the environmental protection effect of a system. The 1990 CAAA established a list of 189 chemicals defined as Hazardous Air Pollutants (HAP) subject to regulation based on two criteria: MACT (Maximum Achievable Control Technology) (equivalent to the results achieved by the best 12 percent of facilities in a source category) and residual risk assessment, which represents the risk to the population from sources of emissions using MACT. Of concern to the coatings industry is that the HAP list includes many of the traditional paint solvents such as xylene, toluene and methyl ethyl ketone. In addition to establishing the HAP list, the CAAA also requires states to develop plans to control VOC emissions under the non-attainment provisions. Implementation of all of the provisions of the act is expected to take several years while regulations are written.

In Germany, solvents are divided in three classes depending on their ecological and toxicological activity. The alcohols used in water-borne coatings belong to the less damaging class III, the glycol ethers and amines by contrast are in class II or even class I [301,102].

The purification of spent air to reduce odor is of increasing significance. A prerequisite for this is the establishment of odor threshold levels by olfactometry.

17.4 Waste - Recycling - Effluent

A number of regulations including the Clean Water Act (CWA) of 1987, the Resource Conservation and Recovery Act (RCRA) of 1980, the Comprehensive Environmental Response, Compensation and Liability Act (CERCLA or "superfund") of 1980 and the Superfund Amendments and Reauthorization Act (SARA) of 1986 address the management of solid waste and liquid effluent. The CWA gives the EPA the authority to reduce or eliminate pollution of navigable waterways and to improve the quality of surface and ground water. RCRA established "cradle to grave" responsibility for the management of hazardous waste. Provisions cover incineration, solid waste disposal, sewage sludge treatment and reporting requirements. CERCLA imposes a tax on chemical raw materials to establish a trust fund for the remediation of abandoned or uncontrolled hazardous waste sites. The act established approved methods for the cleanup of hazardous sites and mandates community "right to know" provisions for extremely hazardous substances.

In Germany the treatment of waste is regulated by provisions that require the paint producer and user to describe the residues created in terms of their chemical and physical properties. Further regulated are the transportation of wastes, the documentation of waste disposal with confirmation of the disposal permit by responsible authorities, the prohibition on mixing of wastes and the requirement to use state of the art disposal techniques for specific wastes [303,304,305,306].

According to the regulations, paint sludges in the future must be disposed off thermally. Since there is inadequate incineration capacity for the near and long term, this implies a need to minimize the creation of paint sludges. Further regulations require firms to minimize wastes if a reduction in water pollution can thereby be achieved.

The recycling of paint coagulants and paint sludges has been intensively investigated by various parties [307,308]. The fundamental results are the following: Binders and pigments are separated from the coagulate to be reintroduced as raw materials [305]. This procedure requires a separate coagulation depending on the type of paint. Other processes separate the used paint binders from the coagulate by membrane technology [309]. These processes are easier to use with solvent coatings than with water-borne coatings. The equipment and energy requirements should not be underestimated. With 2-component coatings, the product must be completely hardened after drying. A subsequent grinding allows the use as filler. Pyrolysis and incineration processes are also discussed [309].

In addition to optimal application techniques, integrated processes to avoid paint sludges are viewed as especially promising [305]. Thus the overspray is captured and recirculated. There are process developments for solvent and water-borne coatings. In practice, there are often failures in color tint consistency and deficient optical properties; this is true especially with high quality requirements, as are present, for example, in the automotive industry. In addition to condensing the paint-contaminated booth water, the process of ultrafiltration is also used. Problems with changes in color tint are thereby resolved [304]. A corresponding facility for the painting of office furniture based on wet paints, powder coatings and electrodeposition paints functions successfully by this technology. The overspray must by captured, collected and reused as pure according to type as possible.

Recycling from the coagulate and the paint sludge is fundamentally possible, but at the time neither ecologically nor economical realistic. Technically the best process is ultrafiltration, where the overspray is recovered directly from the spray booth water. Utilizing the recycled material must be done at the highest level possible. The process is as follows:

The paint is separated from the water in the ultrafiltration facility to which the booth water is continually circulated. The process can be controlled so that the recovered water-borne coating has the same solids content or same viscosity as the virgin material. The recycled paint can be fed directly to the spray gun. A buffer container is necessary when color tint changes result or when the recovered paint is to be analyzed, tinted or adjusted. The recovered water (permeate) is returned to the spray booth. Realistically, the process of ultrafiltration is suited only for water-borne coatings.

Other recovery processes for water-borne coatings function by a continuous electrophoretic deposition of the overspray from the spray booth irrigation water, work-up of the paint coagulate and reintroduction of the booth water [308].

Tied closely to the use of water-borne coatings is the problem of aqueous effluent. Waste water regulations increase the restrictions on the introduction of dangerous substances into the ground water by requiring the use of state-of-the-art technology. The "dangerous substances" are compiled in a constantly revised list by the federal environmental office. In the future all enterprises discharging waste water containing dangerous substances as defined in the Waste Water Source Regulation must satisfy the requirements for state-of-the-art technology.

18 Economic Considerations

In many cases today, water-borne coatings are in many cases still viewed as more expensive and thus less economical than solvent-containing systems. The costs of a coatings system depend on

- the higher research expenditures for relatively new systems
- the initially smaller production and charging scale
- the incompatibility of water-reducible coatings with conventional solvent-containing products and the corresponding higher cleaning expenditures for facilities or the construction of separate facilities for water-borne coatings
- the storage of water-borne coatings under cover at temperatures above freezing, possibly requiring heating.

The costs of an environmentally-friendly coatings system, as represented by the water-borne systems, are however also dependent on the political-ecological environment. If governmental or authoritative dictates require the recovery of the solvents emitted from solvent-containing products and their incineration, and that incineration is subject to carbon dioxide specifications, and if the disposal of solvent-containing paint residues significantly influences the cost situation, then a comparison of the costs between solvent-containing coatings and water-borne coatings turns out quite differently, namely in favor of the water-borne coatings. Since in the near term such a political development in the industrial states of the world must be assumed, a conversion to water-borne coatings should be considered for new investments even when, at the present, solvents can not be avoided for technical reasons.

In using water-borne coatings however, the direct cost saving in comparison with solvent-containing coatings must be considered: The construction requirements including facilities costs for spent air purification, thermal incineration of the spent air, which is tied to high energy costs, absorption processes and their upkeep or the installation and maintenance of biological water treatment disappear. These installations for maintaining air purity are also not free of interruption with the risk of shutting down a paint facility and a loss of production. In working with water-borne coatings, fire prevention measures can be reduced; expenditures for occupational hygiene, such as preventative examinations and the use and furnishing of personal protective measures, are less extensive. The use of water-borne coatings almost always serves to improve the image of the enterprise and, as a measure to protect the environment, can also be used for competitive purposes [310].

Table 32 shows the growth in quantities and value of different coatings systems in the US for the period 1985 – 1991. Compared with solvent-containing coatings (growth 1 – 2% per annum), water-borne coatings have enjoyed above-average growth in both quantities and value. Water-based coatings make up 72%

by volume of all architectural coatings. In 1991 the non-architectural consumption of environmentally-friendly coatings (high-solids, water-borne, powders etc.) was 2 billion pounds worth 3,5 billion US-Dollars, and increasing about 5% per annum.

Table 32 Annual US Paint Sales of Various Coatings Systems (Quantity and Value)[1]

		Architectural Coatings	Product Coatings OEM	Special Purpose Coatings	Total
1985	quantity MM gallons[2]	477	343	138	958
	value MM US-Dollars	3810	3568	1718	
1988	quantity MM gallons[2]	531	364	159	1054
	value MM US-Dollars	4372	4063	2281	10,716
1989	quantity MM gallons[2]	551	357	167	1069.9
	value MM US-Dollars	4713	4236	2484	11,321
1991	quantity MM gallons[2]	539	321	179	1229.8
	value MM US-Dollars	4882	3977	2849	12,923

[1] US Department of Commerce, Current Industrial Rept. 1989, 1990 and 1991, No M28F-12
[2] (1 gallon = 3.785 l)

19 Index of Tables

20 Index of Figures

21 Literature

[1] R. Schaaf, Current Informations of the German Paint Institute „Lack im Gespräch" (1983) No 4, H. Haagen, „Lack im Gesprach" (1894) No 10, 1–12

[2] Voluntary Obligation of the German Paintmakers Association, Frankfurt 1983; „Lack im Gespräch" (1991) No 24

[3] Technical Guidelines for Preserving the Purity of the Air 1986 (TA Luft, Germany)

[4] Production of different coatings systems in Germany/West, Farbe + Lack, 97 (1991) No 7, 626

[5] W. Grüber, D. Stoye, H.-D. Zagefka and H.-J. Zech, Der Lichtbogen (Company's Publication of Hüls AG, Marl) (1983)

[6] J.H. Hildebrand and R.L. Scott, The Solubility of Non-Electrolytes. 3. Ed., Reinhold, New York 1950

[7] C.M. Hansen and A. Beerbower, in Kirk-Othmer, Supp. Vol. p. 889

[8] J.R. Erickson, J. Coatings Techn. 48 (1976) No 620, 58

[9] B.D. Meyer, Oberfläche-Surface, 19 (1978) No 4, 77

[10] A.G. North, J.L. Orpwood and R. Little, J. Oil Col. Chem. Assoc. 59 (1976), 9

[11] L. Serwe, I-Lack 52 (1984) No 7, 262

[12] R.H.E. Munn, J. Oil Col. Chem. Assoc. 74 (1991) No 2, 46

[13] Ch.R. Martens, Waterborne Coatings, Van Nostrand Reinhold Co., New York 1991

[14] R. Laible, Environmentally-friendly Coatings Systems for Industrial Usage, Expert-Verlag, Ehringen 1989

[15] H. Kittel, Lehrbuch der Lacke und Beschichtungen Vol.I, part 1, W.A. Colomb, Berlin-Oberschwandorf 1974

[16] H. Kittel, Lehrbuch der Lacke und Beschichtungen Vol.I, part 3, W.A. Colomb, Berlin-Oberschwandorf 1974

[17] H.-J. Adler, J.W.Th. Lichtenbelt and A.J. Renvers, Farbe + Lack 97 (1991), 103

[18] W. Weger, Pitture e Vernici 66 (1990) No 9, 25–38

[19] A. McLean, Paint Manufacture (1969) No 1 and 5, 45 and 37

[20] P. Oldring and H.C. Hayward, Resins for Surface Coatings Vol.3, SITA Technology, London 1987

[21] P.A. Reitano and T. Szurgyilo, Modern Paint and Coatings (1990) No 7, 44–47

[22] U. Nagorny, Phänomen Farbe 10 (1990) No 9, 45

[23] J.J. Engel and Th. J. Byerley, J. Coatings Techn. 57 (1985) No 723, 29

[24] R.G. Vance, N.H. Morris and C.M. Olson, J. Coatings Techn. 63 (1991) No 802, 47

[25] K. Angelmayer and G. Merten, Phänomen Farbe 10 (1990) No 2, 37

[26] K. Angelmayer, G. Merten and R. Awad, Kunststoffberater (1991) No 3, 20

[27] J.N. Koral and J.C. Petropoulos, J. Paint Techn. 38 (1966) No 501, 610

[28] L.A. Rutter, Adhäsion (1974) No 6, 178

[29] R.C. Wilson, XX. FATIPEC-Congress Nice (1990), 155

[30] Ch.U.Winchester, Proc. 18th Waterborne, High-Solids & Powd. Coat. Symp., New Orleans (1991), 367–396

[31] L. Neevil, Handbook of Epoxy Resins, McGraw Hill 1964
[32] K. Allewelt and W. Göthling, XX. FATIPEC-Congress Nice (1990), 145
[33] H.J. Streitberger, F. Heinrich, T. Brücken and K. Arlt, J. Oil Col. Chem. Assoc. 73 (1990) No 11, 454–458, 464
[34] U. Biethan, J. Dörffel and D. Stoye, Defazet 29 (1975), 447
[35] J. Dörffel and U. Biethan, Farbe + Lack 82 (1976) No 11, 1017
[36] D. Stoye, Double Liaison (1981) No 307, 120
[37] K.H. Albers, A.W. McCollum and A.E. Blood, J. Paint Techn. 47 (1975) No 608, 71
[38] K.L. Payne, F.N. Jones and L.W. Brandenburger, J. Coatings Techn. 57 (1985) No 723, 35
[39] M.R. Olson, J.M. Larson and F.N. Jones, J. Coatings Techn. 55 (1983) No 699, 45
[40] T. Misev, F.N. Jones and S. Gopalakrishnan, J. Coatings Techn. 57 (1985) No 721, 73
[41] H.-G. Stahl, J. Schwindt, K. Nachtkamp and K. Hoehne, Bayer AG, European Patent EP 0140 323 (1983)
[42] D. Stoye, Coating 20 (1987) No 6, 223
[43] Ghanshyan, Pigment and Resin Technology (1990) No 6, 4; W. Freitag, JOT (1992) No 3, 40–44
[44] H. Kittel, Lehrbuch der Lacke und Beschichtungen Vol.I, part 1, 325 ff, W.A. Colomb, Berlin-Oberschwandorf 1974
[45] K. Weigel, Der Farbenchemiker 67 (1965) No 4, 1
[46] G.F. Detrick and E.W. Lewandowski, Resin Review 24 (1974) No 3, 3
[47] A.E. Fischer, Paint, Oil and Col. J. 148 (1965), 999
[48] R. Arnoldus and R.L. Adolphs, Polymers Paint Colour J. 181 (1991) No 10, 405–409,418
[49] E.P. Cruz, Eurocoat 4 (1991), 204–215
[50] A.J. Tortorello and M.A. Kinsella, J. Coatings Techn. 55 (1983) No 697, 29–38
[51] R.E. Tirpak and P.H. Markusch, J. Coatings Techn. 58 (1986) No 738, 49–54
[52] L. Maempel, Adhäsion (1988) No 5, 14–18
[53] W. Cohnen, I-Lack 59 (1991) No 5, 150–151
[54] P.L. Jansse, Polymers Paint Colour J. 181 (1991) No 7, 398, 400, 404
[55] R. Arnoldus, Surface Coatings, Waterborne Coat. (1990) No 3, 179–198
[56] W. Kubitza, Farbe + Lack 97 (1991) No 3, 201–106
[57] W. Kubitza, Pitture e Vernici 67 (1991) No 9, 35–54
[58] W.H.M. Nieuwenhuis, Publication Koninklijke Shell Laboratorium, Amsterdam (Shell Research N.V.), 2–9
[59] F. Bagda, Farbe + Lack 96 (1990) No 12, 934–938
[60] H. Mayer, Farbe + Lack 97 (1991) No 10, 867–870
[61] O. Wagner, Farbe + Lack 97 (1991) No 2, 109–113
[62] M.A. McArthur, Polymers Paint Colour J. 181 (1991) No 4, 164–166
[63] H. Affeldt and G. Koppey, Pitture e Vernici 67 (1991) No 5, 27–40
[64] H.J. Luthardt, Farbe + Lack 87 (1981) No 6, 456–460
[65] A.S. Doyle, Polymers Paint Colour J. 181 (1991) No 6, 44–46
[66] M.E. Woods, Modern Paint and Coating 65 (1975) No 9, 40–47

[67] M.E. Woods, Defazet 30 (1976) No 5, 213–219
[68] A. Jones and L. Campey, J. Coatings Techn. 56 (1984) No 713, 69–72
[69] M.R.C. Gerstenberger and D.K. Kruse, Farbe + Lack 90 (1984) No 7, 563–568
[70] P.M. Grant, J. Coatings Techn. 53(1981) No 677, 33–38
[71] Eastman Chemical Co., Kingsport TN/USA, Publication M-207A (1991)
[72] W. Appelt, Farbe + Lack 96(1990) No 3, 200–201 and J. Oil Col. Chem. Assoc. 73 (1990) No 5, 200–201
[73] R.A. Heckman, Modern Paint and Coating 76 (1986) No 6, 36–42
[74] P.L. Bartlett, Seifen Fette Öle Wachse 113 (1987) No 13, 437–442
[75] W.J. Blank, J. Coatings Techn. 61 (1989) No 777, 119–128
[76] G.L. Burnside, G.E.F. Brewer and G.G. Strosberg, J. Paint Techn. 41 (1969) No 534, 431–437
[77] S.T. Eckersley and A. Rudin, J. Coatings Techn. 62 (1990) No 780, 89
[78] K. Weinmann, Farbe + Lack 93(1987) No 6, 447
[79] J.W. Nicholson, J. Oil Col. Chem. Assoc. 72 (1989) No 12, 475
[80] M. D. Andrews, J. Paint Techn. 46 (1974) No 598, 40–48
[81] H. Dörr and F. Holzinger, Publication „Kronos Titandioxid in Emulsion Paints", Leverkusen 1989
[82] D.M. Wyatt, American Laboratory (1983)
[83] A. Mercurio et al., J. Oil Col. Chem. Assoc. 65 (1982), 227
[84] K.L. Hoy, J. Paint Techn. 45 (1973) No 579, 51
[85] A.J. De Fusco, Coalescing Solvents for Architectural and Industrial Waterborne Coatings, Symposium, New Orleans 1988
[86] K.H. Wallhäußer and W. Fink, Farbe + Lack 91 (1985) No 4, 277 and 91 (1985) No 5, 397
[87] J.W. Gillat, J. Oil Col. Chem. Assoc. 74 (1992) No 9, 324
[88] P.K. Cooke et al., J. Coatings Techn. 63 (1991) No 796, 33
[89] z.B. Hycon, „Keimindikatoren für Flüssigkeiten", Biotest AG, D-6072 Dreieich
[90] J.W. Gilant, J. Oil Col. Chem. Assoc. 74 (1991) No 6, 197
[91] H. Kittel, Lehrbuch der Lacke und Beschichtungen, Vol.II/III, Oberschwandorf 1974
[92] J. E. Hall, J. Coatings Techn. 59 (1987) No 749, 51
[93] P.J. Moles, Polymers Paint Colour J. 178 (1988) No 4209, 154
[94] E. Schinski, Farbe + Lack 91 (1985) No 11, 1019
[95] Rhone Poulenc-Report, Pigment and Resin Techn. 16 (1987) No 4, 7
[96] R.N. Hildred, Farbe + Lack 96 (1990) No 11, 857–859
[97] J.H. Bielemann et al., The Application of Urethane Based Polymeric Thickeners in Aqueous Coating Systems, Lecture, Göteborg 1985
[98] J.E. Glass et al., J. Oil Col. Chem. Assoc. 67 (1984) No 10, 256
[99] S. Le Sota, J. Coatings Techn. 61 (1989) No 777, 135
[100] B. Richey, J. Coatings Techn. 63 (1991) No 798, 31
[101] J.H. Bielemann, Polymers Paint Colour J. 181 (1991) No 4283, 268
[102] P.R. Howard, E.L. Leasure, S.T. Rosier and E.J. Schaller, J. Coatings Techn. 64 (1992) No 804, 87–94

[103] R. Craft, Modern Paint and Coatings 81 (1991) No 3, 38
[104] W. Roelle, Fillers in Water-Borne Coatings, Lecture, Technische Akademie Wuppertal 1991
[105] R.F. Conley, J. Paint Techn. 46 (1974) No 594, 51
[106] K. Meguro and K. Esumi, J. Coatings Techn. 62 (1990) No 786, 69
[107] Bayer AG, Inorganic Pigments for Paints and Coatings - Testing
[108] W. Heilen et al., Coating 20 (1987) No 9, 338
[109] P. Kuschmir, J. Coatings Techn. 59 (1987) No 744, 75
[110] H.F. Fink, W. Heilen, O. Klocker and G. Koerner, Goldschmidt informiert... 1/89 (1989) No 66, 9-21
[111] W. Heilen et al., Coating 20 (1987) No 10, 376
[112] S.J. Storfer et al., J. Coatings Techn. 60 (1988) No 761, 37
[113] R. Zimmerman, Farbe + Lack 89 (1983) No 7, 499
[114] J. H. Bielemann, Farbe + Lack 94 (1988) No 6, 434
[115] E. Talos et al., XVIII. FATIPEC-Congress, Vol. 1/A (1986), 211
[116] P. von Zitzewitz, Water-borne Alkyd Paints, Lecture, Technische Akademie Wuppertal 1990
[117] K. Dören, Corrosion Protection With Water-Borne Emulsion Paints, Lecture, Technische Akademie Wuppertal 1991
[118] H. Koßmann, BASF Publication 11/87, Aqueous air-drying anticorrosion coatings
[119] W. Funke, Farbe + Lack 87 (1981) No 9, 787
[120] H. Leidheiser, J. Coatings Techn. 53 (1981) No 678, 29
[121] New England Society of Coatings Technology, J. Coatings Techn. 54 (1982) No 684, 63
[122] G. Reinhard, Progress in Organic Coatings 18 (1990), 123
[123] R.W. Flynn, J. Protective Coatings and Linings 6 (1989) No 10, 47
[124] E.J. Schaller, J. Paint Techn. 40 (1968) No 525, 433
[125] F. Holzinger and H. Dörr, XVIII. FATIPEC-Congress Vol. 1/B (1986), 327
[126] Ullmann's Encyclopedia of Industrial Chemistry, Vol. A 18, Paints and Coatings, Weinheim 1991
[127] J.G. Balfour, J. Oil Col. Chem. Assoc. 73 (1990) No 6, 225
[128] J.E. Hall and V.R. Pedersen, Surface Coatings Australia 25 (1988) No 4, 12
[129] K.A. Haagenson, American Paint and Coatings J. 72 (1988) No 43, 89
[130] R. Anwari et al., J. Coatings Techn. 62 (1990) No 786, 43
[131] B. Grange et al., Farbe + Lack 93 (1987) No 10, 808
[132] P. Kresse, Farbe + Lack 97 (1991) No 5, 399
[133] H. Luginsland, XVIII. FATIPEC-Congress, Vol. 3 (1986), 55
[134] J. Schmelzer, XIX. FATIPEC-Congress, Vol. 1 (1988), 287
[135] A. Brisson and A. Haber, J. Coatings Techn. 63 (1991) No 794, 59
[136] J. Schmelzer, XX. FATIPEC-Congress Nice (1990), 246-254
[137] L.A. Simpson, Australian OCCA Proceedings and News 20 (1983) No 5, 6
[138] T. Entwistle and S.J. Gill, Pigmentation of waterborne industrial finishes, Publication D 8724 GC of Tioxide International, London (1975)

[139] J.A. Gonzales-Gomez and A. Bartelt, Farbe + Lack 97 (1991) No 2, 97–102
[140] O. Lückert, Pigment- und Füllstofftabellen, 4. Edition, Hannover 1989
[141] Glasurit Handbook „Lacke und Farben", C.R. Vincentz Verlag, Hannover 1984
[142] J. Boxall, Paint and Resin (1990) No 4, 10
[143] M.A. Jackson, J. Protective Coatings and Linings 7 (1991) No 4, 54
[144] P.J. Gardner et al., J. Oil Col. Chem. Assoc. 73 (1990) No 1, 16
[145] S. Turgoose, Polymers Paint and Col. J. 178 (1988) No 4208, 108
[146] J.A. Burkill and J.E.O. Mayne, J. Oil Col. Chem. Assoc. 71 (1988) No 9, 273
[147] A. Bittner, J. Coatings Techn. 61 (1989) No 777, 111
[148] A. Bittner, J. Oil Col. Chem. Assoc. 71 (1988) No 4, 97
[149] O. Leblanc, J. Oil Col. Chem. Assoc. 73 (1990) No 6, 231
[150] H. Wienand and W. Ostertag, Farbe + Lack 88 (1982) No 3, 183
[151] B.P.F. Goldie, J. Oil Col. Chem. Assoc. 71 (1988) No 9, 257
[152] R.S. Hallcoop, Industrial Corrosion (1983) No 8, 18
[153] G. Sugerman and S.J. Monte, Modern Paint and Coating 78 (1988) No 6, 50
[154] L.B. Cohen and S. Eichentopf, Farbe + Lack 94 (1988) No 10, 816
[155] E.V. Carter and R.D. Laundon, J. Oil Col. Chem. Assoc. 73 (1990) No 1, 7
[156] W. Funke, Farbe + Lack 89 (1983) No 2, 86
[157] H. Magdanz, K. Berger and G. Schumann, Farbe + Lack 75 (1969) No 3, 221–235
[158] A. Revillon and Ch. Aiguillon, XX. FATIPEC-Congress Nice (1990), 41–52
[159] J. Spauwen, J. Oil Col. Chem. Assoc. 71 (1988) No 2, 47–49
[160] E.T. Turpin, J. Paint Techn. 47 (1975) No 602, 40–46
[161] W. Siegert, Polymers Paint Colour J. 180 (1990) No 9, 487
[162] Polymeric Dispersions – Handling, Transportation, Storage; Publication of Hüls AG, Marl 1987
[163] S.H. Ashcroft and J.R. Mackness, J. Oil Col. Chem. Assoc. 74 (1991) No 9, 340
[164] P. Toepke, Farbe + Lack 97 (1991) No 1, 38
[165] W. Triebel and P. Hofmann, Farbe + Lack 98 (1992) No 1, 39
[166] G. Kolb and K. Ott, Farbe + Lack 98 (1992) No 2, 125
[167] D.P. Roelofsen, Farbe + Lack 97 (1991) No 3, 235
[168] Publication of Mirodur Company
[169] W. Masing, Handbuch der Qualitätssicherung, 2. Edition, Carl Hanser Verlag, München, Wien 1988
[170] A. Lisson, Qualität: Die Herausforderung, Erfahrungen und Perspektiven, Springer Verlag, Berlin, Heidelberg/Verlag TÜV Rheinland, Köln 1987
[171] J.P. Bläsing, Statistische Qualitätskontrolle, New Edition, Gesellschaft für Management und Technologie AG, St. Gallen 1989
[172] B. Busch, I-Lack 57 (1989) No 2, 55–59
[173] Korrosionsschutz durch Beschichtungen und Überzüge auf Metallen, Weka Fachverlag GmbH, Kissing 1991
[174] W. Beicht und K. Feist, JOT 31 (1991) No 10, 49
[175] JOT Special Edition „Marktübersicht '92" (1992), 71–74
[176] D.W. Glaser, J. Paint Techn. 46 (1974) No 592, 57–64
[177] Anon, JOT 31 (1991) No 11, 50

[178] W. Hater, I-Lack 59 (1991) No 5, 156
[179] J. Kresse, JOT 31 (1991) No 3, 34
[180] K.H. Adams, I-Lack 58 (1990) No 5, 169
[181] G. Blümlhuber, I-Lack 58 (1990) No 5, 175
[182] K. König, I-Lack 59 (1991) No 10, 315
[183] M. Müller, Korrosion 21 (1990) No 4, 183
[184] M. Schröder, Bautenschutz/Bausanierung 13 (1990) No 5, 20
[185] H. Kittel, Lehrbuch der Lack und Beschichtungen, Vol.V, Verlag W.A. Colomb, Oberschwandorf 1977
[186] R. Hantschke, Das Deutsche Malerblatt (1987) No 3, 185
[187] W. Funke and G. Handloser, Farbe + Lack 85 (1979) No 11, 916–919
[188] W. Stelzel, Kunstharznachrichten 24 (1987), 20, and 25 (1988), 4
[189] K. Zimmerschied, Kunstharznachrichten 25 (1988), 7
[190] W. Michel, Das Deutsche Malerblatt (1988) No 4, 269
[191] Das Deutsche Malerblatt (1985) No 3, 215
[192] M. Gebauer, Lecture "Practical Experiences with the Application of Water-Borne Coatings in the Automotive Industry", Haus der Technik, Essen 1988
[193] H. Luderer, Lecture "Modern Applications of Coatings – Elektrostatic and Automation", Haus der Technik, Essen 1988
[194] K. Stahlschmidt, I-Lack 54 (1986) No 9, 375
[195] B. Busch, Lecture "Water-Borne Coatings – Their Application Characteristics With Regard to Industrial Usage", Haus der Technik, Essen 1988
[196] K.D. Ledwoch, I-Lack 59 (1991) No 7, 222
[197] W. Kleber, Water-Borne Coatings and Electrostatical Spraying, 5. Deutsches Wasserlack-Symposium, Bad Nauheim 1990
[198] E. Strauss, Jahrbuch Oberflächentechnik (1988), 395–417
[199] J.R. Taylor, L. Tasker and P.J. Smedley, Defazet 20 (1966) No 1, 12–13
[200] B. Busch, Farbe + Lack 96 (1990) No 5, 331
[201] L. Kühn, I-Lack 35 (1967) No 12, 502–514
[202] G.L. Burnside, G.G. Strosberg and G.E.F. Brewer, Paint and Varnish Prod. (1965) 53–66
[203] S.W. Gloyer, D.P. Hart and R.E. Cutford, Official Digest (1965), 113–128
[204] W. Göring, B. Ancykowski and H. Noack, Farbe + Lack 75 (1969) No 4, 327–336
[205] A.M. Usmari, Organic Coatings, p. 271, Elsevier Publ. Co., New York 1990
[206] D. Rohe, Chemische Industrie (1990) No 5, 88
[207] H. Dinger and G. Wagner, Metalloberfläche 44 (1990) No 8, 392
[208] P.E. Pierce, J. Coatings Techn. 53 (1981) No 672, 52–67
[209] M.S. El-Asser, J.W. Vanderhoff, A. Humayun, C.C. Ho und M.F. Abdel-Bary, J. Coatings Techn. 56 (1984) No 713, 37–42
[210] J. Grey, J. Paint Techn. 41 (1969) No 528, 59–63
[211] H. Heber, Farbe + Lack 97 (1991) No 6, 489
[212] D. Engel, I-Lack 59 (1991) No 12, 403
[213] Anon, JOT 31 (1991) No 12, 28
[214] W.S. Hall, J. Coatings Techn. 52 (1980) No 663, 72

[215] B. Pfeiffer and J.W. Schultze, Bänder Bleche Rohre (1990) No 10, 163
[216] E. Groß, Pulver + Lack 3 (1980) No 5, 416
[217] H. Fink, Plaste und Kautschuk 7 (1980) No 6, 349
[218] H.J. Drexler and B. Biallas, I-Lack 58 (1990) No 6, 208-210
[219] R.F. Eaton and F.G. Willeboordse, J. Coatings Techn. 52 (1980) No 660, 63-70
[220] T. Imai and K. Tsubouchi, J. Coatings Techn. 52 (1980) No 666, 71-78
[221] L.O. Korum, J. Oil Col. Chem. Assoc. 63 (1980), 103-123
[222] J.J. Stratta, P.W. Dillon and R.H. Kemp, J. Coatings Techn. 50 (1978) No 647, 39-47
[223] A.L. Rocklin, J. Coatings Techn. 50 (1978) No 646, 46-55
[224] A.L. Rocklin and D.C. Bonner, J. Coatings Techn. 52 (1980) No 670, 27-36
[225] W. Göldner, I-Lack 58 (1990) No 6, 205
[226] D. Rasch, Lecture "Practical Experiences with Water-Borne Coatings", Technische Akademie, Wuppertal 1991
[227] A. Heinemann, Coating 24 (1991) No 3, 96
[228] E. Beck, E. Keil and M. Lokai, Farbe + Lack 98 (1992) No 3, 165
[229] P.W. Dillon, J. Coatings Techn. 49 (1977) No 634, 38
[230] A. Bischoff, Oberfläche-Surface 20 (1979) No 3, 54-57
[231] R.S. Bailey, Modern Paints and Coatings 68 (1978), 43-46
[232] M. Yate, Polym. Paint Colour J. 168 (1978) No 3969, 128
[233] K.-H. Adams, Fachtagung Wasserlack '91, Proceedings, München 1991
[234] P. Herbold, Fachtagung Wasserlack '91, Proceedings, München 1991
[235] F. Clancy, European Surface Treatment (1991) No 7, 44-47
[236] G. Dingler, Oberfläche + JOT (1990) No 1, 24-27
[237] K. Weigel, I-Lack 34 (1966) No 10, 442-444
[238] Audi, Die Lackiererei im Audi-Werk Ingolstadt, Produktion 16.05.1991, No 20, 23-24
[239] Opel, Die Lackiererei im Opelwerk, Produktion, 16.05.91, No 20, 20
[240] Anon, I-Lack 59 (1991) No 2, 59-66
[241] R. Kraus, I-Lack 59 (1991) No 10, 318-321
[242] M. Gebauer, Lecture "Water-Borne Paints and Coatings Systems", Technische Akademie Wuppertal, Symposium Handbook 1990
[243] F. Rey and R. Gould, JOT 31 (1991) No 7, 50-51
[244] K.-H. Frangen, I-Lack 34 (1966) No 1, 4-17
[245] W. Maisch, I-Lack 34 (1966) No 12, 515-521
[246] W. Maisch, I-Lack 33 (1965) No 9, 299-313
[247] A. Matting and H.-D. Steffens, Defazet 20 (1966) No 11, 509-517
[248] E.P. Miller and L.L. Spiller, Official Digest 37 (1965) No 589, 117-132
[249] B. Queng, I-Lack 38 (1970) No 1, 5-9
[250] H.U. Schenck and J. Stoelting, J. Oil Col. Chem. Assoc. 63 (1980), 482-491
[251] M. Ramasri, G.S. Srinivasa Rao, P.S. Sampathkumaran and M.M. Sirshalkar, J. Coatings Techn. 61 (1989) No 777, 129-134
[252] K. Kingberger, Phänomen Farbe 11 (1991) No 6, 22
[253] K. Kingberger, I-Lack 59 (1991) No 10, 332-333

[254] V. Stevens and P. Griggs, XX.FATIPEC-Congress Nice (1990), 333 – 336
[255] C.W. Metzger, R. Laible and H. Ternberger, I-Lack 60 (1992) No 1, 5 – 6
[256] E. Hess, H. Böhnke and R. Wolf, XX.FATIPEC-Congress Nice (1990), 287 – 291
[257] H. Rauch-Puntigam, Österreich. Chemie-Zeitschrift (1983) No 1, 7 – 12
[258] H. Rauch-Puntigam, Oberfläche-Surface 17 (1976) No 7, 141 – 145
[259] H. Metzdorf, I-Lack 58 (1990) No 4, 127 – 130
[260] Anon, Oberfläche + JOT (1990) No 4, 62
[261] Anon, Oberfläche + JOT (1991) No 3, 26
[262] L. Tasker and J.R. Taylor, J. Oil Col. Chem. Assoc. 48 (1965) No 2, 121 – 149
[263] K.-R. Hefendehl, I-Lack 52 (1984) No 6, 205 – 208
[264] W. Burckhardt and H.J. Luthardt, J. Oil Col. Chem. Assoc. 62 (1979), 375 – 385
[265] Anon, JOT 31 (1991) No 12, 22, 25 – 26
[266] G. Thöresz, Oberfläche + JOT (1991) No 3, 28 – 30
[267] Boge/München, Oberfläche + JOT (1990) No 12, 22 – 26
[268] E.J. Percy and F. Nouwens, J. Oil Col. Chem. Assoc. 62 (1979), 392 – 400
[269] A.H. Peters and E. Kramer, I-Lack 52 (1984) No 7, 248 – 250
[270] S.A. Stachowiak and P.G. Kooijmans, XX.FATIPEC-Congress Nice (1990), 95 – 100
[271] Th. Katsibas/Reichhold-Albert, German Patent DP 2 054 468 (1970)
[272] E. Landwehr, Farbe + Lack 96 (1990) No 12, 923 – 933
[273] W. Burckhardt, I-Lack 46 (1978) No 8, 265 – 269
[274] R. Bartscherer, I-Lack 59 (1991) No 6, 184 – 188
[275] H. Schwarze, Coating 24 (1991) No 8, 311 – 313
[276] H. Cabos, Coating 24 (1991) No 9, 346 – 348
[277] Anon, Oberfläche + JOT (1990) No 12, 15
[278] K.H. Weinert, Fachbroschüre Oberflächentechnik (1986) No 1, 19 – 21
[279] Anon, JOT 31 (1991) No 11, 76 – 80
[280] J. Fichtner, Oberfläche + JOT (1990) No 5, 26 – 28
[281] B. Gürer, J. Sprenger and T. Wurster, Oberfläche + JOT (1990) No 8, 20 – 22
[282] Anon, JOT 31 (1991) No 12, 30 – 32
[283] A. Keiler, I-Lack 58 (1990) No 9, 327 – 330
[284] K. Heberlein, I-Lack 58 (1990) No 5, 165 – 168
[285] G. Schmitz and D. Emmrich, I-Lack 57 (1989) No 2, 60 – 63
[286] G. Walleter, I-Lack 58 (1990) No 10, 375 – 379
[287] Ch.M. Winchester, J. Coatings Techn. 63 (1991) No 803, 47 – 53
[288] K. Dickerhof, I-Lack 60 (1992) No 1, 7 – 13
[289] C. Godau, I-Lack 58 (1990) No 4, 134 – 138
[290] Anon, I-Lack 60 (1992) No 2, 49 – 56
[291] J.M. Akkerman, Das Deutsche Malerblatt (1992) No 2, 30 – 37
[292] T. Neuteboom, Pitture e Vernici 66 (1990) No 4, 18 – 24
[293] R.H.E. Munn, J. Oil Col. Chem. Assoc. 73 (1990) No 7, 286 – 290
[294] S. Knödler, Das Papier 35 (1981) No 10A, 74 – 79
[295] C.W. Patterson, J. Oil Col. Chem. Assoc. 73 (1990) No 7, 290 – 294

[296] W.Kern, J. Oil Col. Chem. Assoc. 74 (1991), No 12, 436-445
[297] L.J. Culver and P.M. Grant, American Paint & Coatings (1979)
[298] L.J. Culver, Modern Paint and Coatings (1981)
[299] W. Hansen, Farbe + Lack 87 (1981) No 7, 551-556
[300] W. Lohrer, JOT 32 (1992) No 4, 108-119
[301] K.-H. Weinert, I-Lack 47 (1979) No 6, 216
[302] J. Sarbach and G. Schlumpf, Oberfläche + JOT (1991) No 3, 18-20
[303] E. Mink, Oberfläche + JOT (1991) No 3, 22-26
[304] J. Diebold, Oberfläche + JOT (1990) No 11, 857-859
[305] H. Deuster, I-Lack 58 (1990) No 9, 325-326
[306] K. Panzer, I-Lack 58 (1990) No 9, 323-324
[307] P. Murr, JOT 32 (1992) No 3, 34-39
[308] J. Halbartschlager, JOT 32 (1992), No 3, 28-33
[309] H. Upgang, JOT 31 (1991) No 6, 52-57
[310] BASF, Handelsblatt (1990) No 28, 14

22 Key word index